When Time Breaks Down

ARTHUR T. WINFREE

When Time Breaks Down

The Three-Dimensional Dynamics

of Electrochemical Waves and

Cardiac Arrhythmias

PRINCETON UNIVERSITY PRESS

Published by Princeton University Press, 41 William Street,
Princeton, New Jersey 08540
In the United Kingdom: Princeton University Press, Guildford, Surrey

Library of Congress Cataloging in Publication Data will be
found on the last printed page of this book

ISBN 0-691-08443-2
ISBN 0-691-02402-2 (pbk.)

This book has been composed in Linotron Sabon and Gill Sans

Clothbound editions of Princeton University Press books
are printed on acid-free paper, and binding materials are
chosen for strength and durability. Paperbacks, although satisfactory
for personal collections, are not usually suitable for library rebinding

Printed in the United States of America by Princeton University Press
Princeton, New Jersey

DESIGNED BY LAURY A. EGAN

謹以此書致吾妻吉雲！

——設有她的幫助
本書將尚未付梓。

CONTENTS

PART II
Rotating Waves

PART III
Organizing Centers

CHAPTER 8
Patterns of Timing in Three-Dimensional Space

CHAPTER 9
A Bestiary of Organizing Centers

PART IV
Postlude

CHAPTER 10
What's Next?

ACKNOWLEDGMENTS

Writing was done while on leave from Purdue University, thanks to support from the Oxford University Centre for Mathematical Biology, the Guggenheim Foundation, the MacArthur Foundation, Los Alamos National Laboratory, the Institute for Natural Philosophy, and the University of California Institute for Nonlinear Science. The National Science Foundation has generously supported these investigations since 1965, particularly with grants from the Division of Chemistry's Chemical Dynamics Program and the Division of Cellular Biosciences' Cellular Physiology Program after this book was outlined in 1981. Noncomputerized artwork was executed by Roy Allen, Dennis Davidson (airbrush), and Ji-Yun Winfree. I thank Tim Poston for computerizing Figure A.7, Melvin Prueitt and Steven Strogatz for help in programming some figures in Part III, and the University of California / Los Alamos National Laboratory for access to Cray computer facilities for construction of the jacket illustration and many figures in chapters 8 and 9. Steven Strogatz suggested the main title. Leon Glass, John Tyson, Raymond Ideker, José Jalife, and Michael Guevara purged many foolish assertions. For the perseverance to finish after many changes of direction, I owe special thanks to torrents of photons on Beer Can Island in the Gulf of Mexico and on Solana Beach in Southern California. The patience and understanding of my wife Ji-Yun remain indispensable.

28 February 1986

SOURCES AND CREDITS

Sources of previously published figures are acknowledged in the captions; further information is listed below in accordance with the requirements of the copyright holders.

1.5, 1.6 Reprinted (1.5) or adapted (1.6) with permission from W. Taylor et al., "Critical Pulses of Anisomycin Drive the Circadian Oscillator in Gonyaulax toward Its Singularity," *Journal of Comparative Physiology* 148 (1982):11-25, figs. 6 and 7.

2.5 Reprinted with permission from J. Roelandt et al., "Sudden Death during Longterm Ambulatory Monitoring," *European Heart Journal* 5 (1984):7-20, fig. 4.

3.2, 3.4 Adapted from *Lectures in Applied Mathematics* (1981) "Peculiarities in the Impulse Response of Pacemaker Neurons," A. T. Winfree, vol. 19, figs. 1 and 3, by permission of the American Mathematical Society.

4.6 Reprinted with permission from E. N. Best, "Null Space in the Hodgkin-Huxley Equations," *Biophysical Journal* 27 (1979):87-104, fig. 13.

4.8 Adapted with permission from J. Jalife and C. Antzelevitch, "Phase Resetting and Annihilation of Pacemaker Activity in Cardiac Tissue," *Science* 206, fig. 2, p. 696, 9 November 1979.

4.9a Adapted with permission of the authors and the American Physiological Society from S. V. Reiner and C. Antzelevitch, "Phase-Resetting and Annihilation in a Mathematical Model of the Sinus Node," *American Journal of Physiology* 249, H1143-53, fig. 9.

5.4 Adapted from M. A. Allessie et al., "Circus Movement in Rabbit Atrial Muscle as a Mechanism of Tachycardia: II. The Role of Nonuniform Recovery of Excitability," *Circulation Research* 39 (1976):168-177, fig. 2, by permission of the American Heart Association, Inc., and the senior author.

5.5 Adapted with permission from the senior author and the American College of Cardiology, from E. Downar et al., "On-Line Epicardial Mapping of Intraoperative Ventricular Arrythmias," *Journal of the American College of Cardiology* 4 (1984):703-714, fig. 8.

5.6 Adapted with permission from A. L. Wit et al., "Electrophysiologic Mapping to Determine the Mechanism of Experimental Ventricular

Tachycardia Initiated by Premature Impulse," *American Journal of Cardiology* 49 (1982):166-185, fig. 3.

5.8 Reprinted from M. J. Janse et al., "Flow of Injury Currents and Patterns of Excitation during Early Ventricular Arrhythmias in Acute Regional Myocardial Ischemia in Isolated Hearts," *Circulation Research* 47 (1980):151-165, fig. 8 and 11, by permission of the American Heart Association, Inc., and the senior author.

6.7 Adapted from H.J.J. Wellens et al., "Ventricular Fibrillation Occurring on Arousal from Sleep by Auditory Stimuli, *Circulation* 46 (1972):661-664, fig. 1, by permission of the American Heart Association, Inc., and the authors.

7.2 Reprinted by permission from M. A. Allessie et al., "Experimental Evaluation of Moe's Multiple Wavelet Hypothesis of Atrial Fibrillation," in *Cardiac Electrophysiology and Arrhythmias*, edited by D. P. Zipes and J. Jalife (Orlando, Fla.: Grune & Stratton, 1985), fig. 9.

7.3 Reprinted with permission of the *Journal of Chemical Education* from A. T. Winfree 1984a (61:661).

7.9 Reprinted from P. C. Newell, "Attraction and Adhesion in the Slime Mold Dictyostelium," in *Fungal Differentiation*, edited by J. Smith (New York: Marcel Dekker, 1985), fig. 3, by courtesy of Marcel Dekker, Inc.

7.17(p. 183) Reprinted with permission from S. C. Müller, T. Plesser, and B. Hess, "The Structure of the Core of the Spiral Wave in the Belousov-Zhabotinsky Reagent," *Science* 230, fig. 2 left, p. 662, 8 November 1985. Copyright 1985 by the AAAS.

8.4, 8.5 Reprinted with permission from A. T. Winfree, "Scroll-Shaped Waves of Chemical Activity in Three Dimensions," *Science* 181, fig. 1, p. 937, 7 September 1973. Copyright 1973 by the AAAS.

8.6 Reprinted with permission from A. T. Winfree and S. H. Strogatz, "Singular Filaments Organize Chemical Waves in Three Dimensions: 1. Geometrically Simple Waves," *Physica* 8D (1983):35-49, fig. 11.

8.7, 8.8, 8.9, 8.10, 8.12, 8.15, 9.5, 9.10, 9.17 Reprinted with permission from A. T. Winfree, "Wavefront Geometry in Excitable Media: Organizing Centers," *Physica* 12D (1984):321-332, figs. 5-8, 11, 13, 14, 18.

9.9, 9.15 Reprinted with permission from A. T. Winfree and S. H. Strogatz, "Singular Filaments Organize Chemical Waves in Three Dimensions: 3. Knotted Waves," *Physica* 9D (1983):333-345, fig. 7.

9.14 Reprinted with permission from A. V. Panfilov and A. T. Winfree, "Twisted Scroll Rings in Active Three-Dimensional Media," *Physica* (1985):323-330, fig.8.

When Time Breaks Down

The soul always worries and tries to make order from
that which doesn't take to being ordered.
—An elderly widow of central Nepal
(quoted in *CoEvolution Quarterly*, Spring 1980)

Attempts to decipher the unique patterns that make living things so different from the nonliving must have been among the earliest exercises of mankind's speculative imagination. It is a natural preoccupation: we are living organisms, we hunt and are hunted by living organisms, we grow living organisms to eat. Yet we don't understand enough about living organisms or even enough about ourselves. Being curious, we play Sherlock Holmes, looking, asking questions, and trying things out. We struggle in our imaginations to weave a close copy of the patterns we find, ever hopeful of apprehending a covert orderliness that is not entirely our own invention. Naturally, the resulting fabric is mostly made of the stuff of dreams. What we think we apprehend seems always to squirm and metamorphose as though it too were a living thing.

This book is mostly about patterns involving time. Long before Darwin, time was already a special concern of biologists: because of the central drama of evolution by which matter came to be alive, because of the abbreviated replay of that drama in the unfolding of each individual organism from a single cell, and because of the curious ways behavior is structured in time. Memory, heartbeat, the menstrual cycle, and even the life cycle remind us of eddies and whirlpools in the fabric woven of living matter, space, and time.

Biologists have traditionally been preoccupied with time, and some mathematically minded biologists have thought deeply about it. J.B.S. Haldane was one of the greatest of these, both as a man and as a mathematical theorist of the origin of life and its evolutionary development. In a 1927 collection of short popular essays, *Possible Worlds*, Haldane entertained his readers by speculating how differently many things would appear if time ran not in a one-dimensional endless line but on a forked path like the branching tree of evolution, or if it were two dimensional, more like a broad river than a thread, or not endless but closed in a ring.

"Now my suspicion is that the universe is not only queerer than we suppose," he said, "but queerer than we *can* suppose."

The idea that time might, in some respects, run in a circle is nothing new, of course. It is built into Indo-Hellenic philosophies and was entertained by Plato, Pythagoras, and Aristotle, the first biologist in the modern tradition. A circular concept of time is the usual bias of pre-industrial societies, to whom the future seems mainly a repeat of the past. In our own society, it remains a subject of frequent personal wishfulness and occasional good poetry, as well as an inspiration for abundant science fantasy featuring the paradoxes of looped cause and effect.

In many respects biological time does run in a circle, just as day and night do on a rotating planet. There are surprising paradoxes implicit in this queer situation: not the ones celebrated by science-fiction writers, but others that are equally unfamiliar. They are unfamiliar largely because biological and chemical oscillators have only been subjected to acute scrutiny since the early 1960s. This lapse might seem strange, given that reliable repetition of patterns that have been found to work is an obvious organizational principle in the kinds of material that lend themselves to the evolution of life. But living materials are diverse in ways that often defy the mathematics evolved for doing physics and thus in those terms seem imprecise and unanalyzable. Biologists have recently recovered from this illusion in many ways. The way of particular pertinence here is the recognition that there are modes of mathematics—of "reasoning with symbols"—other than the ones that make living organisms look imprecise. The topological mode offers special promise. It is this mode—indeed, one tiny theorem in one part of this mode—that is celebrated here. Take note that this book remains tightly focused on experimental biology and chemistry. There will be no explicit mathematics. There is almost none behind the scenes, either; the kinds of topology involved really boil down to little more than geometric intuition applied with patient tenacity.

There is now a science of temporal anatomy. Its subject is behavior, especially periodic behavior and its pathologies. This subject matter includes the circadian rhythms that mesh our daily ups and downs with each other and with the immutable cycle of day and night—and the pathologies of circadian rhythms that bring insomnia and perhaps depression without any chemical, physical, organic, or structural disease. It includes the robust choreography of muscular contractions in running, breathing, the heartbeat—and the arrhythmias (irregularities) of the heartbeat that undermine so much otherwise promising vitality. It includes even stranger processes: whirlpools of electrical activity in heart muscle and the rotating organelles recently discovered in chemical soups that were originally intended to re-create in a test tube the energy-producing reactions of living cells.

Part I of this book is concerned with biological rhythms as purely temporal patterns without any spatial organization. The focus is on the organization in time that we sense in circadian rhythms and its susceptibility to shattering by a flash of light near midnight; and on the quicker rhythms of the mammalian heartbeat, their electrical signatures, and the susceptibility of the rhythm to lethal disruption by an unfortunately timed stimulus. The emphasis is on singular arrhythmias of perfectly normal and healthy clocks, abruptly induced by an ephemeral, ostensibly innocuous stimulus.

Part II considers spatially organized clocks in living systems and in a more readily comprehensible biochemical analogue. We will examine the electrical organization in space and time that constitutes the normal heartbeat, its conversion to lethal arrhythmia by a singular stimulus. The vortex-like spatial pattern of this arrhythmia is revealed by electrical measurements in heart muscle and by visual inspection of analogous systems.

Part III studies the spatially and temporally rhythmic self-organization of surprisingly life-like objects found in a nonliving chemical caricature of excitable tissues. The focus here is on the geometry of spatially and temporally periodic excitation, whatever its incarnation may be, including the three-dimensional thickness of dying heart muscle, with computer graphics paving the way for recognition of spatially organized arrhythmias.

Throughout we will discover again and again in a surprising diversity of contexts the same paradoxical entity: a motionless, timeless organizing center called a phase singularity. This is a place where an otherwise pervasive rhythm fades into ambiguity—like the South Pole, where the 24 hourly time zones converge and the Sun merely circles along the horizon.

To pursue this geographical metaphor, a global map can be colored in time zones ranging smoothly from purple (perhaps at the Greenwich meridian) to violet (through Moscow) to blue (Lhasa) to turquoise (Tokyo) to green (Midway Island) to yellow (Pitcairn Island) to orange (New Orleans) to red (Rio de Janeiro) and back to violet (at Greenwich).

Only eight time zones are named here but in principle the globe could be painted in the same pattern using 360 hues for 360 successive phases of the daily cycle. Around the equator labeling hues advance smoothly, as the wave of sunrise races uninterruptedly westward. Only at two points—the Poles[1]—must the coloring inevitably become confused. These

---

[1] To be strictly correct, only the time zones of convention always go to the Poles. The real time zones—the loci of instantaneous sunrise or sunset—do so only on the two equinox days of each year. On any other day the real time zones end along a circle that seasonally expands from each Pole to the Arctic or Antarctic Circle. Similarly, we will find later that the locus of confused timing in other rhythmic systems is commonly a finite closed ring rather than a mere point.

phase singularities will recur in more abstract time zone maps throughout the coming chapters, each occasion signaling some singular disruption of rhythmicity in a new context. Their inevitability follows from simple geometry; we can pass from one context to another, despite diverse underlying mechanisms, without losing our bearings. However, it is not exactly geometry and geometric intuition that clear our path, but really topology and topological intuition.

Topology is a mode of "reasoning with symbols" about continua such as velocities, colors, temperatures, surfaces, rings, braids, angles, shapes, and about continuous arrangements of one continuum in another, such as the temperature distribution on the Earth's surface. The "topological" properties of such continua are unaffected by gradual distortions such as stretching, bending, cooling, reddening, and so on. Topological properties of objects are usually discrete: being an inside or an outside, having a certain number of vertices or holes, being knotted or linked in a certain way. Such properties can be discussed in a precise way without having to measure curvatures or distances or solve differential equations.

The academic study of topology dates back at least to Leonard Euler in Konigsburg, but its vigorous pursuit in connection with dynamics (time) is only as old as the twentieth century. It already has enjoyed vogues of enthusiastic popularization among experimental scientists. For example, René Thom's sevenfold classification of the "catastrophes" [Thom 1972] of nonoscillating dynamical systems was celebrated in the *London Times* Sunday Magazine and in *Scientific American*. Christopher Zeeman motivated many to see biological and psychological analogies [Zeeman 1977]; engineers found the same phenomena in structural failures of materials under load, and physicists came to a clearer vision of complex optical phenomena [Thompson and Hunt 1977]. The newer mathematics of "chaos" has now reached the *New York Times* Sunday Magazine and the *Economist*. It too is largely topological in flavor; current enthusiasms touch on weather prediction, stock market fluctuations, and cardiac arrhythmias.

There are some amazing theorems in topology. Thus far we have exploited only one of them, concerning rings (appropriate in connection with periodic time), in a diversity of contexts. There are comparable theorems about a two-clock system—for example, a bilaterally symmetric animal with independent clocks in each hemisphere of the brain, with states not on a ring but on the space associated with two rings: the surface of a doughnut. With every mechanical or dynamical system there is an associated continuum of possible states, often conveniently represented as some strange geometrical object (see Chapter 10). Perplexing behaviors of the system then correspond to topological invariants of that continuum.

Spheres, for example, are natural to discussions of direction in three dimensions. The orientation of fibers in a polymer such as DNA packed in a cell's nucleus is aptly described for many purposes by a point on a sphere. Consider these topological inferences about the two-dimensional sphere, cast in terms of the Earth's surface [from Tucker and Bailey 1950]:

(1) There must somewhere be a place with dead calm wind: the wind cannot be active horizontally everywhere at the same time. (The point(s) where it is momentarily calm typically skitter about with arbitrary speed, not limited by the speed of light.)

(2) If there is no such becalmed point in the whole Northern Hemisphere (unlikely as that may be), then at that moment you can find somewhere on the Equator a place where the wind is blowing in whatever compass direction you like. This inference actually follows from the theorem that will be overworked in this book, substituting direction for phase and hemisphere for disk.

(3) There is a place somewhere in the Northern Hemisphere such that, at the antipodal place in the Southern Hemisphere, the barometric pressure and temperature are exactly the same, to arbitrarily many decimal places.

These are facts specifically about a sphere; they are not all guaranteed, for example, in the doughnut-shaped or ring-shaped worlds of science-fiction writer Larry Niven. The topology of a sphere is essential, but the apparent frivolity is not an essential aspect; the facts are guaranteed just as irrevocably in matters of practical engineering concern.

But nothing is fun to discuss in the absence of background. A context can be built by reexamining the familiar phase singularities of circadian clocks, known since the late 1960s. From there the discussion will turn to analogous but less familiar peculiarities of neural pacemakers, on a time scale a million times shorter, and thence to visible apparitions of the same phenomena in cardiac tissues, the retina of the eye, and chemically reacting fluids. In all cases but the first (circadian rhythms), mechanisms are reasonably well understood and quite diverse. We will dwell not on mechanisms but on the common principles of timing that organize them all. The first chapter will not even elaborate much on the principles, but rather will review the main phenomena to be sought next in electrophysiological and biochemical contexts. Principles will be developed from the basics when those contexts are encountered in Chapter 2. (If Chapter 1 seems too stark, the interested reader can see Winfree 1980 and 1986a for extended discussion.) We now turn to a review of the phase singularities of circadian clocks, for as Isaac Newton declared in another age of discovery about ancient and venerable clockworks, "In learning the sciences, examples are of more value than precepts."

PART I

Arrhythmia

CHAPTER I

Cellular Rhythms and

Their Singularities

In "The Final Problem" the master sleuth, Sherlock Holmes, confides just before his death, "Of late I have been tempted to look into the problems furnished by Nature, rather than those more superficial ones for which our artificial state of society is responsible." Had Holmes developed that theme earlier in life, he might have made the professional acquaintance of Gregor Mendel and Charles Darwin (already old men during Holmes's youth) and of younger contemporaries who also approached the riddles posed by Nature with the same joyful spirit of shrewd detection, achieving memorable solutions to the mysteries of the day. One that has proved curiously refractory to solution, however—perhaps it is the one that was earmarked for Holmes—tantalized Darwin persistently enough to prompt an entire book, *On the Power of Movement in Plants* (1880). This problem is the mystery of biological clocks, still unsolved in terms of almost any of the traditional modes of biological explanation: in terms of the selective value of internal circadian rhythmicity, in terms of its evolutionary history, in terms of its physiological mechanism, in terms of its biochemistry. The situation is still much as Holmes declared in the Adventure of Silver Blaze:

> We are suffering from a plethora of surmise, conjecture, and hypothesis. The difficulty is to detach the framework of fact—of absolute, undeniable fact—from the embellishments of theorists and reporters. Then, having established ourselves upon this sound basis, it is our duty to see what inferences may be drawn and which are the special points upon which the whole mystery turns.

In 1986 it is not yet possible to do what Holmes suggests, except by backing off from the specific problem of circadian rhythms to contemplate biological clocks in a broader context, including neural and cardiac pacemakers and even biochemical oscillators. From that perspective, we begin to see some recurring patterns, not unlike geological fault lines organizing the ostensible diversity of local landscapes.

Despite the variety of mechanisms that underlies rhythmic timing and its control in living organisms, several modest generalizations stand out. For example, any spontaneously rhythmic mechanism ("oscillator") can be expected to lock on to the nearby period of a strong enough rhythmic influence. (Linear oscillators are important exceptions but they are not often encountered in chemistry or biology.) This ability of most nonlinear oscillators to entrain or synchronize rests on their time-dependent sensitivity: exposed to some standard disturbance beginning at different times in the cycle, there will be different phase shifts inflicted. Almost a quarter-century ago this principle was first articulated clearly enough to make sense of entrainment phenomena encountered by neurobiologists [Perkel et al. 1964] and to anticipate phenomena that, from a less analytical perspective, seemed paradoxical: for example, the heartbeat can be accelerated by an inhibitory stimulus, made to cycle faster by more frequent inhibition, and slowed down by increasing the rate of stimuli that individually accelerate it. These common events became understandable in the early 1960s when physiologists first paid analytical attention to phase resetting in nonlinear oscillators.

About the same time similar ideas were found useful to students of the long-period physiological rhythms that mesh any organism's behavior to the night/day alternation of its environment. A few crucial conceptual innovations and a great deal of careful experimental work in circadian physiology clarified apparently universal patterns of phase resetting and breakdown of rhythmicity in response to a stimulus. In particular, a "vulnerable phase," when a stimulus of exact size can cause circadian timing to break down or at least become unpredictable (see below), was discovered in the circadian cycle. This isolated stimulus was named "critical annihilating stimulus," and the arrhythmic center in the pattern of timing got the name "phase singularity"; it is indeed "the special point upon which the whole mystery turns."

Circadian physiology became, and until recently remained, the preeminent vehicle for refinement of ideas about timing and its adjustment by phase-resetting stimuli. The circadian era in the development of such principles seems nearly finished, as greater opportunities now abound for investigators to decipher the neural and molecular mechanisms of circadian rhythmicity. But mechanisms of the more rapid oscillation of pacemaker membranes are already within the grasp of neurobiologists; the time is ripe in that field to resume inquiry into organizational principles of timing. Rhythmic timing is especially intriguing in the context of spatially distributed pacemakers such as the stomach, intestine, kidney, and uterus—and the heart, which is the prime example in this book.

This book mostly skips over the circadian era of discovery about clocks, their resetting, and their breakdown (adequately portrayed elsewhere [Winfree 1980; Winfree 1986a]); instead it attempts to continue the story in its original neurobiological context, extending the original applications from artificially isolated pacemaker neurons to pacemakers embedded in their natural context: in a tissue of oscillatory and excitable cells.

Stimuli applied in this more realistically geometrical context elicit geometrical rearrangements of timing. In fact, the same rearrangements occur even in excitable tissues that lack the local property of spontaneous oscillation. Such tissues—the muscular wall of the heart, for example—are susceptible to patterns of timing that persist perniciously, once instigated, by directing a wave along a closed path to circulate forever. At the center of the vortex is a point of arrhythmia closely related to the arrhythmic singular point in patterns of phase resetting.

These vortices have a three-dimensional structure that is not yet easy to examine in living heart muscle. But they can be created and destroyed in a chemically reacting medium that lends itself more conveniently to study of the novel principles involved in this kind of arrhythmia. They will be easier to recognize in complex biological media after they are first perceived clearly in a more idealized experimental medium.

This chemical medium—the Belousov-Zhabotinsky reaction—was discovered in 1951 during biochemical studies of the Krebs cycle [Winfree 1984a; Belousov 1985]. It may be the first completely understandable laboratory example of pattern formation in a chemical system that involves nothing more than chemical reaction and molecular diffusion. (Diffusion is the random migration of molecules: see Glossary.) The possibility of pattern formation by reaction/diffusion mechanisms was discovered, also about 1951, by a theorist—the same Alan Turing who initiated the art of digital computation [Turing 1952]. Despite ponderous theoretical development in a literature of thousands of technical papers over three decades, the idea has lacked a single clear laboratory illustration. The Russian reaction now provides one. But its antics turn out to resemble nothing foreseen in the thirty years devoted to the subject by theoretical chemists and biologists, as will be seen in later chapters.

A thumbnail review of the pertinent phenomena discovered in the past two decades by phase-resetting experiments with circadian clocks follows. These experiments were motivated by explicitly mathematical theory that posed questions and prescribed experimental means to get a yes/ no answer. When the necessary experiments were implemented, their main results sometimes turned out to match those foreseen. These successful predictions (a rarity in theoretical biology) were derived not from what might have seemed the pertinent considerations of physiology or natural selection, but from abstract notions of continuity (see Glossary) and nonlinear dynamics. Their vindication thus reveals nothing about questions of keen interest from the viewpoint of mechanism. Because conjecture about the actual (unknown) mechanisms of circadian rhythmicity was never really involved, the same conceptual scaffolding, reinforced and enriched by knowledge of actual mechanisms of electrophysiological pacemakers, might equally well support comparable inquiries into the organization of timing in contexts where the known mechanisms are almost too complex to grasp. Subsequent chapters will suggest ways to begin.

Resetting a Circadian Clock

Theoretical scaffolding is momentarily dispensed with here in order to portray the patterns of phase control observed in circadian systems. For readers curious about engineering principles, the scaffolding from which these experiments were built is compactly accessible in the Appendix: it summarizes the essential concepts from Winfree 1980. A less concentrated development of the necessary ideas follows this chapter: the scaffolding will be reassembled from the ground up starting in Chapter 2, where we begin a search for comparable phenomena in neurobiology.

Despite a few determined efforts, circadian rhythms have not yet been detected in procaryotes—that is, in cells whose genomes are diffuse rather than packaged as nuclei and whose mechanisms for deriving chemical energy by oxidation are not packaged as mitochondria. But most other organisms, unicellular or not, do exhibit predictable ups and downs of physiological function and behavior, with an internally determined period near 24 hours. Hence the name (coined by Franz Halberg in 1959) *circa*dian. These spontaneous internal rhythms are normally entrained to lock step with the daily rhythm of environmental temperature, light intensity, and so on. But in conditions of "temporal isolation" these clocks "free-run"; their native period is then evident.

Inquiries into the mechanism of oscillation and its entrainment by external stimuli led to phase-resetting experiments. In a typical phase-resetting experiment, the free-running organism in temporal isolation is exposed to some stimulus protocol beginning at a certain phase of the cycle (the "old phase") and ending after some assigned duration. The subsequent free-running rhythm is then observed. It generally differs from that observed in a control not exposed to the stimulus (or exposed for a sufficiently brief time): the phase has been reset from the old phase to a new phase. The exact way in which it differs, of course, depends on the duration of exposure and generally depends on when in the cycle the stimulus began. A plot of the new phase against the old phase is called a "phase-resetting curve." There are reasons for laboriously conducting

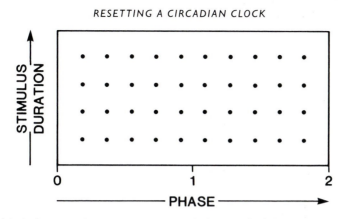

Figure 1.1: A phase-resetting experiment composed of many individual measurements, each employing a single stimulus of some strength or duration (increasing northward) applied at some time (increasing through two full cycles, redundantly) eastward.

such protocols and there are insights to be inferred from the outcome. But let us for the moment bypass these and look at the results naively. The most unvarnished presentation of the raw data is best for such purposes.

First we will plot the times and durations of all the experiments performed. Each experiment will be represented by a dot in a horizontal rectangle whose two dimensions are (to the north) stimulus duration and (to the east) circadian old phase when the stimulus began, measured in units of one circadian period starting from when the organisms were placed in temporal isolation in some standard physiological condition (Figure 1.1).

So much for the experimentalist's input. What about the organism's output? We will plot the time from the end of the stimulus to the monitored circa-daily event (and the next, and the next, etc.) by placing a button to represent each event along a wire calibrated in units of one circadian period, beginning from the moment the stimulus is finished and the organism is therefore free-running again. Each experiment will create one such wire with buttons on it, standing erect above the corresponding point on the horizontal rectangle of stimuli (Figure 1.2).

The First Measured Time Crystal

Figure 1.2 [Winfree 1970ac] shows the first such construction, in which a brief exposure to dim blue light reset the circadian timing of a fruit fly's first act as a sexually mature adult: emergence from the pupal case. Of course, the individual fly emerges only once in a lifetime. But in a population of hundreds, individuals become mature enough to emerge

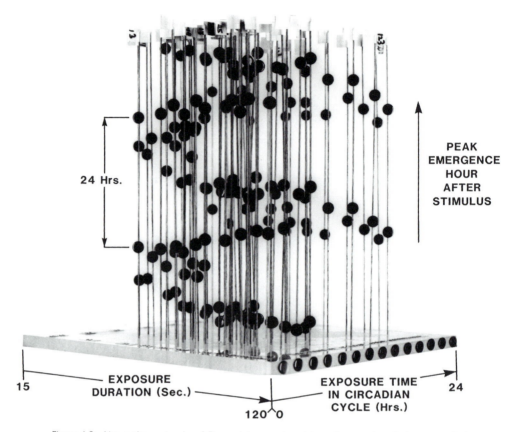

Figure 1.2: Above the rectangle of Figure 1.1, experimental results are plotted along a vertical time axis. Only one full cycle of phase is shown horizontally. Each button locates the middle of a daily emergence peak at an altitude proportional to its time in hours after the stimulus given for that measurement. Collectively the buttons outline a screw-shaped surface. Adapted from Winfree 1970ac.

hour by hour over a span of a week or so, depending on growth conditions. During that week, bursts of emergence activity recur every 24 hours. Their timing betrays the timing of the internal circadian clock common to all individuals of whatever development age. In Figure 1.2 this population rhythm is plotted vertically for each of a few score populations that were each exposed to a single light pulse during otherwise continuous darkness. The plot is rough, but it shows the essential features that have appeared in every subsequent repeat of this measurement [Winfree 1980; Peterson 1980, 1981] with diverse stimuli and diverse organisms (but not yet including Man).

First of all, the results are periodic in two directions. Events recur periodically after the stimulus, giving the plot a repeat in altitude once in every circadian cycle. Second, the results are the same whether the stimulus is given in one cycle or the next (but only one cycle is shown here horizontally), just as long as it is given at the same phase in the cycle,[1] so the plot is periodic to the east. The plot thus consists of unit cells like a crystal, each one cycle wide and one cycle high. It is commonly called a time crystal.

What structure occupies each unit cell of the time crystal? In the first 24 hours above each dot on the horizontal rectangle—ideally above each point, given unlimited lab time—there hovers a single button. Collectively they make up a smooth surface completely hiding the rectangle when viewed from above. What kind of surface? Before examining the actual surface in any detail, imagine what it must resemble in two extreme cases.

Suppose first that the stimulus, regardless of when it is given or for what duration, resets the circadian cycle to a standard phase. Then all the wire and button displays will be the same: the first button floats the same distance above stimulus time (the floor) on every wire, the next one cycle above that, and so on. Each unit cell contains a featureless horizontal plane in this imaginary extreme case. This case is of course implausibly extreme, since there must be some lower limit of stimulus duration before which it cannot have noticeable impact.

Let us then consider in contrast the opposite extreme case in which the stimulus has no effect at all, regardless of duration; this is, of course, the situation with most stimuli, e.g. someone sneezing in a remote country. In this experiment event times are unmoved: each event still occurs at the same hour, measured from the beginning (not from the stimulus). Time from beginning to stimulus (horizontally to the east) plus time from stimulus end to event (upward) is fixed. (This sum leaves out stimulus duration only because duration happened to be short in these particular experiments.) Thus the buttons lie along a downward sloping plane.

The Resetting Surface Has a Singularity

So much for imagination; what about data? With inept choice of stimuli, data in fact resemble one or another of these extremes. Given that experience, a moderate choice can be made so that the results are sensitive to variations in the stimulus in ways that depend on the nature of the circadian mechanism. As Figure 1.2 hints, the measured resetting surface then rises like one turn of a screw in each unit cell. Figure 1.3 shows the

---

[1] Qualifications: (1) The organism's sensitivity to the stimulus must be taken into account. It is common for tenfold dark adaptation to develop in 2–3 days, affecting the impact of light stimuli; (2) The organism may react differently during first half-cycle or so of temporal isolation, if it was previously kept in rather different conditions.

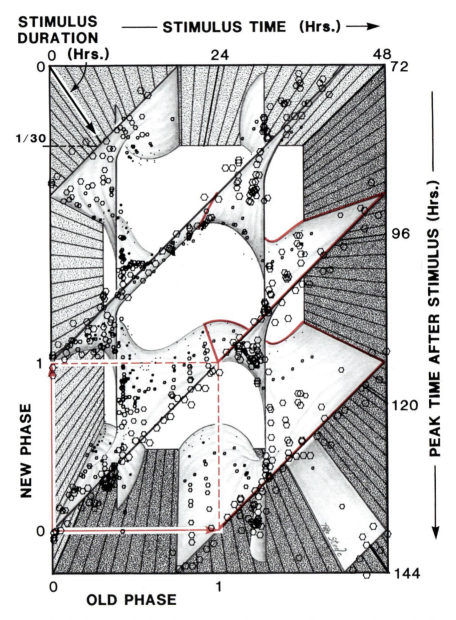

Figure 1.3: Similar to the physical original, this computer plot shows the same experiment as Figure 1.2 supplemented by many more measurements. The screw surface is sketched by hand. Six unit cells are shown: 2 cycles of phase horizontally by 3 days' recording vertically *downward* from 72 to 144 hours after the stimulus. Stimulus magnitude increases from 0 (big hexagonal "dots") to 2 min. (tiny dots) into the background. Adapted from Winfree 1980, where multiples of 24 hours after light-to-dark transfer were called phase 0 as here, but the new phase axis was mislabeled.

same data, supplemented by more experiments to a total of 510, plotted in stereo projection with a smooth surface fitted to the cloud of data points. The coordinates differ a little from Figure 1.2: time is not plotted upward, but rather *downward* from the roof, three cycles after the stimulus, and *two* cycles of stimulus onset time are shown from left to right. In the foreground (arbitrarily brief exposures), the resetting surface is the tilted plane, signifying negligible effect. Where not negligible, a weak stimulus still elicits resetting that follows a wavy downward diagonal. This is called *odd* resetting because every altitude is encountered an odd number of times as stimulus timing is varied from left to right through one full cycle. In the background (stronger stimuli), resetting data follow a wavy horizontal surface akin to the horizontal plane foreseen in the limit of very strong stimuli. This is called *even* resetting because every altitude is encountered an even number of times as stimulus timing is varied from left to right through one full cycle.[2]

In the middle is a screw axis around which the resetting surface leads smoothly from every datum to every other. Such a surface would be awkward to grasp in the hand and is no less difficult to grasp in the imagination. It helps to see a contour map, just as it does in making sense of a ski slope. To make a contour map one cuts horizontal slices of the terrain, starting at the highest hilltop and working down to the lowest stream bed or lake surface. Each slice is a curve or maybe a ring or several rings of land, all at the same elevation. These slices are all dropped onto a common plane: the map paper. On some maps they are color coded for altitude: commonly brown for high elevations, grading into green for valleys. We will also use a color code, but elevations one circadian cycle apart will be given the same color because here elevation means phase in a cycle, just as in coloring the time zones in the Introduction. Thus successive turns of the screw surface are colored alike and all superimpose on the contour map. Or, if you like, we are doing a contour map of only one unit cell in the conventional way except that the highest elevation is colored the same as the lowest. Figure 1.4 is such a map of the surface

---

[2] What we here call "even" resetting is often called "Type 0" resetting because the average slope of the resetting surface is 0 where it merely fluctuates about the horizontal. It is sometimes also called "strong" resetting because it necessarily entails advances or delays (or both) exceeding 1/2 cycle. "Odd" resetting is correspondingly called "Type 1" to remind us that the resetting surface is wobbling about a diagonal of unit slope. It is also sometimes called "weak" because it is compatible with very weak phase shifts and always obtained in the limit of very weak stimuli. But the mean slope is the key characteristic, not the magnitude of phase shifts. The size of phase shifts can be misleading: violent resetting can be part of a Type 1 pattern if the new phase increases steeply with increasing old phase, but then turns around with further increase. The numerical names, though more meaningful, are harder to remember, so we still use "odd" (for Type 1) and "even" (for Type 0). There are other "odd" types (3, -1, etc.) and other "even" types (2, -2, etc.) but they have not yet been encountered in biological experiments.

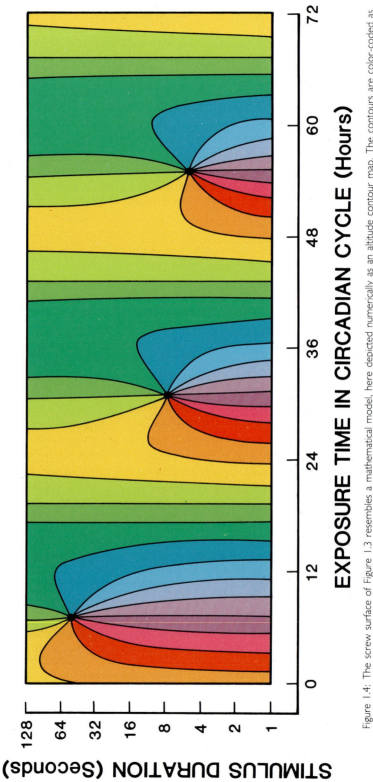

EXPOSURE TIME IN CIRCADIAN CYCLE (Hours)

STIMULUS DURATION (Seconds)

Figure 1.4: The screw surface of Figure 1.3 resembles a mathematical model, here depicted numerically as an altitude contour map. The contours are color-coded as described in the text. Note the orderly convergence of all colors in sequence to the singularities at which circadian arrhythmia occurs.

used to fit the data in Figure 1.3. (Actually that surface was sketched by hand, whereas Figure 1.4 was computed numerically; but the computed surface is indistinguishable by eye from the sketch and fits the data to within ±2 hours, only a little worse than replicate control experiments fit one another. See Winfree 1972b for details.)

Every color contour represents one uniform altitude, or value of reset timing; such contours are accordingly called isochrons. The isochrons are equispaced along the zero-stimulus axis at the south edge of the rectangle. The smooth progression of color through one full cycle at the south edge of the rectangle of stimuli means only that if the stimulus has little impact then the time from stimulus to next event decreases steadily as the stimulus is placed later and later in the cycle. Each isochron touches the south edge, but not all make it to the north edge: large stimuli reset the rhythm to more or less the same phase, regardless of initial phase, leaving out a wide range of contours.

None of this seems the least bit mysterious until we contemplate taking a stroll across this contoured landscape, expecting upon return to home base that we will find ourselves again at the original altitude. Certainly this is the result on any landscape contoured by the U.S. Geodetic Survey; it is also exactly true here on most paths that vary only the timing or only the size of the stimulus or that vary both together, but not too much. In contrast, however, if we vary both in the right way on this phase-resetting landscape, we return to find our altitude one day lower at the original coordinates. One example of "the right way" is a circumnavigation along the borders of the rectangle. This somewhat puzzling "case of the missing day" is encountered along any path that encloses the central singularity, where all timing contours converge. Near that point they converge radially, like the steps of a spiral staircase. This is the axis of the screw surface.

What is the import of this screw axis? There the resetting surface becomes arbitrarily steep: event times cannot be predicted following a stimulus of that special magnitude, if it is given at that special time, the "vulnerable phase" of the fly's circadian time sense. Following stimuli near the screw axis—the phase singularity, as it is called—the fly's circadian clock is nearly turned off. The closer the stimulus is to the critical annihilating stimulus, the less vestige of bias remains in the fly's choice of emergence time or in its reaction to the timing of a subsequent stimulus. The clock mechanism, whatever it is, has not been damaged: any later flash of light provides a cue that restarts the marking of time by circadian intervals and restores the fly's smoothly phase-dependent response to a later stimulus. But in the absence of that cue, flies remain arrhythmic, just as they are if initially conceived and reared in conditions devoid of any time cue. Having experienced the singular stimulus, they reverted to that time-neutral condition.

Singularities in Other Circadian Clocks

Mosquitoes have circadian clocks quite similar to flies', which can be monitored in the adult by recording rest and activity times. Following the singular stimulus, mosquitoes who are normally active at dawn and dusk seem to have turned insomniac: they fly and rest irregularly at all hours, day after day [Peterson 1980; Peterson 1981].

This behavior is not limited to animals with nervous systems. Consider the results of conducting the same protocol while monitoring not the activity of an insect, but the circadian timing of cell division in suspensions of the unicellular alga *Euglena*. In this case [Malinowski et al. 1985] the stimulus was a 3-hour exposure to light at any time in the circadian cycle, of intensity ranging from 0 (in control experiments) to a brightness that saturates the response. The timing of daily cell division was reset in the even pattern following bright exposures (and, of course, in the odd pattern following negligible exposures). Following a stimulus of appropriately intermediate intensity, given near the end of the daily interval of mitoses, cell division in this population switched from daily synchronous bursts to random timing. Uniform exponential growth replaced the former staircase increase of cell numbers. The reason may be that every cell's circadian oscillator was turned off, thus depriving cells of any particular time reference for gating mitosis; or it may only be that each cell returned to the usual circadian cycle at random phase, leaving the population incoherent. The distinction can be made experimentally, but this chore still remains for the future.

A more graphic example is provided by the circadian rhythm of bioluminescence in the marine alga *Gonyaulax polyedra*. This unicellular plant, often responsible for the so-called "red tides," lives in sea water and photosynthesizes in the sunlight. At night, or at 23-hour intervals in temporal isolation, its continuous faint glow (bioluminescence) reaches a maximum roughly tenfold brighter than during its minimum in the opposite half-cycle. You may have seen this luminescence as brilliant blue flashes around your body while swimming in the sea at night. Brief application of an inhibitor of protein synthesis such as anisomycin can reset the timing of this glow to crest at any time of day (and at 23-hour intervals thereafter) [Taylor et al. 1982].

In Figure 1.5 we see a contour map of the resetting surface of *Gonyaulax*, derived from about 500 separate experiments. Because they were all executed before any data were analyzed, stimuli are not optimally placed according to the "singularity trap" protocol [Winfree 1980; Peterson, 1980, 1981]: large flat regions are overpopulated with experiments, and some regions of rapid change are unsampled. But in overall appearance the surface seems much as before: a smooth staircase at small

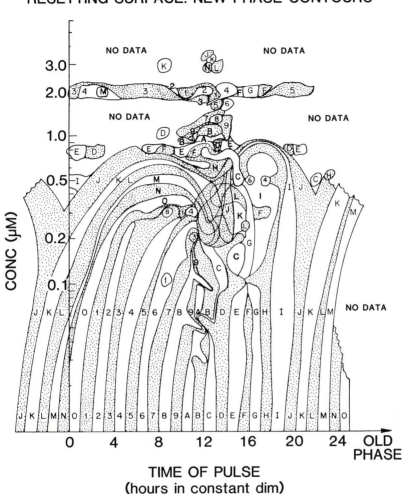

ANISOMYCIN 1h PULSES
RESETTING SURFACE: NEW PHASE CONTOURS

Figure 1.5: Analogous to Figure 1.4 (but not colored), we see here the experimentally measured resetting contours of a different organism's circadian clock. Both contour maps are based on a sampling of about 500 experiments. This contour map [Taylor et al. 1982, with permission] represents the time crystal of *Gonyaulax*, its clock perturbed by a 1-hour pulse of the protein synthesis inhibitor anisomycin at a concentration indicated vertically, at the hour (old phase) indicated horizontally. The resulting new phase can be read by following any region or a same-numbered region to low concentration, where new phase = old phase. Some regions inadvertently left unlabeled in the original publication are labeled here, courtesy of R. Krasnow.

23

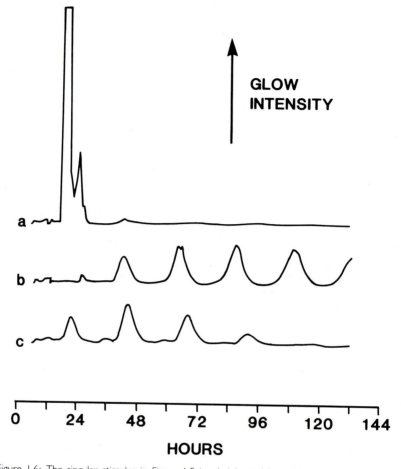

Figure 1.6: The singular stimulus in Figure 1.5 is administered here (a) while bioluminescence is monitored. A conspicuous surge of intensity is followed by days of quiescence. This was not observed after the other stimuli or after the same stimulus applied at other times (e.g. b, c). Adapted from Taylor et al. 1982, Figure 7, with permission.

stimulus size, with the steps converging to a central pole (heavy overlap of contours). Figure 1.6 shows what happens (a mere offset of the peak timing) when most resetting stimuli are administered and, in contrast, when the singular pulse is administered. For reasons unknown there may be a monstrous emission of light several hours later, followed by little or no emission day after day [Taylor et al. 1982]. The cells remain perfectly healthy, but their bioluminescence seems to no longer be organized in a daily pattern.

The Pinwheel Experiment: A Vortex of Light

Figures 1.4 and 1.5 each show a snapshot of an array of many experiments. The experiments were described as though conducted separately, by repeating a standard protocol over and over with slightly different timing. And so they were. But in principle they might have been done all at once. Suppose for example that a film of *Gonyaulax* suspension were poured in a rectangular tray under bright light and gelled. Bright light usually holds circadian clocks at fixed phase; *Gonyaulax* provides no exception. When plunged into darkness, cells resume oscillation. By slowly moving a shadow across the tray from east to west during a 23-hour interval, clocks in north-south columns are started one after the other. An east-west phase gradient then exists, just as in Figure 1.5. By then pulling the shadow southward and sliding it back north, we expose cells simultaneously in east-west rows, the more northerly cells being exposed longest and cells near the south edge seeing almost no light at all. Bright or prolonged light inflicts even resetting on *Gonyaulax*, much as does anisomycin. Thus the entire array of stimuli has been set up in this one experiment. Cells in various parts of the tray will be reset much as indicated by the anisomycin contour map for *Gonyaulax*, or, more clearly, by the light contour map for *Drosophila* (Figure 1.4). Isochron 0 (red) shows which cells will be glowing maximally after 23 hours, again after 46 hours, and so on. Adjacent isochron 1 (reddish-violet) shows where glow was maximal an hour ago. Next is isochron 2, and so on. The glow maximum proceeds as a wave moving from one isochron to the next like a rotating pinwheel, pivoting clockwise about the motionless singularity where the cells have no rhythm. Thus the name "pinwheel experiment."

This rotating maximum is more like a tide than a wave in the usual sense, since nothing is really propagated or conducted from one cell to the next. It is only an optical illusion derived from the orderly pattern of timing across the tray. It rotates in the same period regardless of distance from the pivot and has no characteristic speed. It would not be blocked by killing a swath of cells to cut a "firebreak" across the tray.

But what if adjacent cells were metabolically or electrically coupled? Then activity in the present wavefront would have a chance to activate adjacent cells prematurely, and the wave might really propagate. It might tend to propagate around the initial pivot, because the way is already paved by a circular gradient of timing. In any case the wavefront (isochron) does have an edge at the pivot; something peculiar must occur at that singularity. In Parts II and III we will encounter exactly this situation and discover what happens.

25

Pinwheel Experiments in Diverse Contexts

Pinwheel experiments have now been repeated in many contexts: various circadian rhythms in plants, animals, and fungi [Winfree 1980]; the cat's respiratory oscillation [Paydarfar et al. 1986]; rhythms of nuclear division in micro-organisms [Kauffman and Wille 1975]; the oscillations of energy metabolism in yeast cells [Winfree 1980]; water-transport oscillations in sprouting seedlings [Johnsson and Karlsson 1971; Johnsson 1976]; periodic firing of nerve cells and cardiac pacemakers [Winfree 1981, 1982ab, 1983]; and even an inorganic chemical oscillator [Sevcikova et al. 1982]—but almost always in the awkward piecemeal format of the first pinwheel experiment (see Box 1.A). Nevertheless something like the ideally simple protocol may be enacted spontaneously in heart muscle under conditions that lead to rotating waves and sometimes to sudden death [Winfree 1983]; and something very much like it can be demonstrated in the laboratory in liquid films of the chemical oscillator [Winfree 1985]. The principle involved in wave rotation is implicit in the time crystal, as we saw in its contour map.

The lesson from many pinwheel experiments is that the isochrons do converge in an orderly way to a point. In drawing attention to that unexpected point, the experiment still leaves unresolved the fundamental ambiguity mentioned above. Suppose the experiment is actually performed with such accuracy and in such a noise-free situation that the exact singularity is achieved: the rhythm is annihilated. What does it mean to annihilate a biological rhythm in a multicellular organism or suspension of many cells? Is the clock in each merely paralyzed somewhere along its cycle? Has the circadian mechanism lapsed into a stupor of timelessness? Or has it only lost the former macroscopic coherence of microscopic oscillations that still continue, randomly reset? Or are they perhaps only asynchronously reset, not randomly but rather in a beautiful pattern throughout the three-dimensional bulk of the tissue? In Parts II and III we will find examples of this last possibility in periodically active chemical reactions and in heart muscle.

But returning to the spatially unstructured oscillations thought to underlie circadian rhythmicity, there are essentially two choices for the clock's behavior following a singular annihilating stimulus: either it recovers promptly or it doesn't. In *Drosophila* it doesn't. Cells of the flower *Kalanchoë* appear to behave similarly [Engelmann and Johnsson 1978]: rhythmicity is suppressed following the singular stimulus. In other species, e.g. the cockroach [Wiedenmann 1977], rhythmicity recovers promptly (because the exact singularity was missed slightly and/or physiological fluctuations promptly bumped the circadian system away from its singularity) with unpredictable timing.

Box I.A: A Two-Dimensional Circadian Clock

Drop a spore of the bread mold *Neurospora crassa* onto the glistening surface a sterile culture medium. Within a day it germinates and sends growing filaments in all directions across the moist agar surface at about 1 mm per hour. Kept in the dark at constant temperature, the spreading mycelium uniformly oscillates with a circadian rhythm that can be assayed biochemically or visually by the 21-hour alternation of growth types at the advancing frontier: sometimes making spores, sometimes merely advancing. By transplanting a tiny bit of mycelium from far behind the present frontier to fresh medium, it can be induced to resume this patent alternation too, thus providing a direct biological assay of clock phase at any point in the two-dimensional sheet of circadian clocks [Dharmananda and Feldman 1979].

Can these clocks be reset by chemicals or by light? Certainly. Even-type resetting is easily obtained in *Neurospora* [Nakashima et al. 1981; Dharmananda 1980]. A complete pinwheel experiment can be conducted within a single baking dish by appropriate manipulation of gradients of dim blue light [Winfree and Twaddle 1981]. In this situation a central patch of mycelium is trapped at its singular state within a ring of mycelium oscillating normally, but at all phases of the cycle in succession around the ring. Whether the singular state is inherently stable or even violently unstable, execution of the pinwheel experiment conjures it into the middle of the mycelium and it has no way out. Any subsequent stimulus will move it in one direction or another, but can move it outside the bordering ring of mycelium whose clocks still span the full cycle only if that second stimulus is strong enough to inflict even resetting on the border. The distance and direction of movement directly monitors the effect of that stimulus on the clock.

This sensitive exploration remains incomplete for want of a strain of *Neurospora* that perseveres in steady growth for as long as the 7–10 days required to set up the pinwheel experiment in stable oscillation. A second technical difficulty is the tendency of time zones to spread at the expense of their neighbors in a temporally graded mycelium—presumably more due to lateral growth competition along the frontier than to chemical coupling between adjacent patches of oscillating tissue [Winfree and Twaddle 1981]. If coupling does play a role, further experiments with this system may reveal waves of circadian resetting, including rotating pinwheel waves such as we will see in Part II and Part III in neural and chemical contexts analogous to this.

A Primitive Theory of Circadian Dynamics

In all cases checked, the time crystal contains a screw surface in each unit cell. This result was predicted from each of three independent notions [Winfree 1980]:

(1) It might be supposed that the circadian mechanism involves the continuous interplay of quantities that affect one another's rates of change (e.g. biochemical concentrations, membrane potentials, enzyme activities) and that a stimulus ultimately affects those rates. The interactions may be complex, involving time delays, compartments, and many variables. But if they generate an attracting limit cycle (see the Appendix), then the qualitative theory of differential equations sets limits on the topology of resetting surfaces. To contrive a plausible model that produces anything very different from a time crystal of screw surfaces requires some ingenuity.

(2) Circadian rhythms in all presently convenient experimental systems actually pool the rhythms of many cells. Those cells are individually rhythmic but are not known necessarily to remain synchronous during resetting. It is surprisingly difficult to contrive a vision of the individual cell's circadian mechanism and its susceptibility to disturbances such that the collective rhythm of independent cells would not react qualitatively as above.

(3) All circadian rhythms react to small stimuli in the odd resetting pattern. A great many, tested with sufficiently potent stimuli, react in the even pattern. None has exhibited any other pattern. Given that the phase response changes continuously as stimulus duration or magnitude is changed (except possibly for an isolated combination of starting phase and stimulus size) the qualitative pattern described above is almost inevitable.

So What?

This new phenomenon does not immediately deliver into our hands a nugget of new understanding. If the existence of a phase singularity was not foreseen by most biologists two decades ago, it is only because they had not thought deeply about the implications of already familiar findings. The discovery of phase singularities has done nothing yet to illuminate the evolution, physiology, or molecular mechanisms underlying circadian clocks. The ambiguous interpretation of singular arrhythmia in multicellular tissues only underscores the importance of monitoring the clock in single cells rather than in collectives. In cases where annihilation reflects arrhythmia in every cell (rather than a randomization of phases in the population), it potentially provides a delicate experimental tool for detecting factors that affect the mechanism and perhaps for classifying them according to the stage at which they intervene in the mechanism. For example, all protein synthesis inhibitors might restart a singularly arrhythmic clock at phase 1/3, whereas brief anoxia or other mitochondrial disturbances might dislodge it from equilibrium toward phase 2/3, while intervention through one photoreceptor might initiate rhythms at phase 0, and through another toward phase 1/2. We have at present no other comparably delicate assay for interference with the clock's inner workings; but this one has not yet been exploited (see Box 1.A).

What does any of this mean? Perhaps not much. Our situation is like that of the anonymous industrial consultant whose final report concludes: "We have not succeeded in answering all your problems. The answers that we *have* found only serve to raise a whole new set of questions. In some ways we feel we are as confused as ever, but we believe we are confused on a higher level and about more important things." This is a familiar situation in science. By elucidating the counterintuitive properties of a gyroscope, no one discovered an amendment to Newton's laws of motion or inferred anything about the nature of matter. They were, however, better able to understand and anticipate the strange behavior of gyroscopes and to remedy their ailments. With this moral in mind we now turn to oscillations and arrhythmias in excitable membranes.

CHAPTER 2

The Heart in Life and Death

I was almost tempted to think ... that the motion of
the heart was only to be comprehended by God.
—William Harvey, *De Motu Cordis
et Sanguinis in Animalibus* (1628)

In Chapter 1 we discovered the pinwheel wave, created by a critical stimulus and organized around a phase singularity. But the context was contrived; the pinwheel experiment is a rather artificial format for organizing the results of phase-resetting measurements on separate biological clocks. What if there were some natural context in which phase resetting naturally occured in a "pinwheel experiment" format, with the reset clocks adjacent and interacting with their neighbors? Would the phase singularity become the center and source of a rotating wave? Box 1.A suggested that this may occur in two-dimensional sheets of circadian clocks. In Part II (beginning with Chapter 5), we will discover such a context in the human heart. The consequent disorganization of the heartbeat is an arrhythmia that can lead in a matter of seconds to fibrillation and sudden death. Our objective in Part I, however, is only to establish a base camp for that assault. Chapter 1 reviewed at great speed the main phenomena of phase resetting and phase singularity as first encountered in circadian clocks. In Chapters 3 and 4 they will be encountered again in a more leisurely way in pacemaker membranes, including those of the heart. But first, in this chapter, I attempt to provide a motivating context by a review of the normal heartbeat, its uniform rhythm, and its remarkable vulnerability to abrupt disorganization in spatial patterns that resemble the results of a pinwheel experiment.

An Indispensable Oscillator

The human heart is a qualified guru for training aspirants to understand things rhythmical. It may be the most elaborately engineered clock on this planet. No other human organ has evolved specifically to function as an oscillator under such remorseless selective pressure. There is no backup in the event of a failure. Yet the heart does sometimes fail. The normal coordination of parts in the heart, like the hypnotic synchrony of a good juggler's performance, can come unraveled in a great variety of ways. These ways are called arrhythmias, though few of them really lack all semblance of rhythmicity. Dysrhythmia is more descriptive for most of them: alternative rhythms that the heart may fall into, usually without disastrous consequence in the short term. Your circadian juggling act also has its dysrhythmias; you might feel the consequences of them when you change daily habits, for example, after a switch to daylight saving time and certainly when you experience jet lag or a less expected week of insomnia and daytime drowsiness. These slower, more ponderous melodies instruct us in the principles of temporal organization and disorganization. But instruction is slow in laboratories limited to one data point per day. Learning might be quickened by focusing on the 100,000 times livelier tempo that you can feel in your own pulse. How can the principles gleaned from circadian physiology be applied on the time scale of the heartbeat? What do the arrhythmias of circadian clocks hint at, how can the hints be translated into the language of cardiac physiology, and how might those translations offer new insight into the causes of lethal arrhythmia?

Answers are only recently coming into focus, mostly from mathematically inclined physiologists and physiologically inclined mathematicians in research centers scattered around the world. The answers seem complicated. But curiously it is easier in some respects to comprehend the onset of incipient total disaster—the enigmatic catastrophe called fibrillation—than it is to sort out the diverse mechanisms of lesser arrhythmias. We will strive to appreciate the working of universal principles in the timing of our own heartbeats by exploring one way that a perfectly healthy young heart can instantly and irreversibly lose its formerly flawless coordination. This does happen: the commonest cause of death by electrical shock is fibrillation; and fibrillation can be induced by less artificial stimuli, as when "self-electrocution" occurs by sudden emotion in individuals whose heart has become electrically unstable.

Sudden cardiac death can occur anywhere at any time, unpredictably. If you spend half an hour reading this chapter, you can be confident that hundreds of individuals will meanwhile add to the worldwide statistics of this mishap. There are a thousand cases each day in the United States alone. About one man in five will eventually be stricken down by sudden cardiac death. The shorter average lifespan of men is largely due to this single factor; the greater susceptibility of males—about threefold—can be seen in the proportionately larger number of women in retirement homes. A cardiology symposium proceedings entitled *Sudden Death 1980* begins with the words: "Over the last ten years it has become increasingly obvious that Sudden Death represents the major challenge confronting cardiology in the last part of the twentieth century."

The challenge encompasses matters extending beyond our preoccupation here with the human heart: we seek to understand general principles of biological timing and its susceptibility to abrupt breakdown. Sudden cardiac death may illuminate these principles in that it can occur as a response of perfectly healthy heart muscle to a stimulus (e.g. electrical shock) that somehow—we will try to understand how—abruptly randomizes the timing of rhythmic contractions all over the heart. The new conceptual ingredient here is "all over": we must now come to grips with spatial organization of timing. The pinwheel-experiment interpretation of circadian arrhythmia will provide our bridge from time to space. But in addition to a new conceptual challenge, we now confront the additional challenge of coming to terms with an overwhelming profusion of complicated facts: the immensely detailed richness of medical physiology as it bears on the workings and failures of the human heart.

In the diversity of its material components and the complexity of their interactions, the heart resembles a mechanical engine with dozens of built-in counterbalancing electrical governing systems. The hearts of diverse mammals studied in the laboratory differ distinctively from human hearts, and even individual human hearts vary significantly in anatomy and behavior. Any one heart, in normal operation, is continually adapting to changes in its host's psychological and emotional state, physical workload, and chemical balance. The variety of abnormal behaviors and the corresponding massiveness of cardiology textbooks reveal what an intricate balancing act underlies the appearance of simplicity in regular beating. The epigraph of this chapter catches the frequent mood of every student of this subject up to the present day. To embark on anything like an adequate description is to abandon all hope of writing plainly in a short space.

A well-known investigator summarized our plight one hundred years ago: "The principal difficulty of your case," remarked Sherlock Holmes in his didactic fashion (in "The Matter of the Naval Treaty"), "lay in the fact of there being too much evidence. What was vital was overlaid and hidden by what was irrelevant. Of all the facts that were presented to us, we had to pick just those that we deemed to be essential, and then piece them together in their order so as to reconstruct this remarkable chain of events." Let us follow Holmes's exhortation to simplify ruthlessly.

The principles of phase resetting provide a shortcut through this maze. Though they were discovered by biologists working with 24-hour cycles, they are not tied specifically to the mechanisms of circadian clocks: no one knows those mechanisms, except to guess that some are "attracting-cycle" (alias "limit-cycle") oscillators (see Appendix). The pacemaker of the heart actually has an attracting cycle, with only the additional bewildering complication that it also has parts with a spatial arrangement. Thus we may have a ready-made, laboratory-tested shortcut in hand already. The shortcut provides safe conduct as far as a cardiological analogue of the pinwheel experiment in spontaneously rhythmic tissue such as the sinus node. Extension to the more interesting case of rhythmically active heart muscle (atrial and especially ventricular muscle), which is not spontaneous, is mediated through a mathematical analogy in the Appendix. I believe that phase singularities can be the seeds of rotating waves even in healthy heart muscle. And we will see that rotating waves are the immediate antecedents of the total chaos called fibrillation.

The Normal Human Heartbeat

Before dwelling on irregularities of the heartbeat, it is useful to review its normal structure and function and to see how physiologists might analyze a beating heart in the abstract style that permits both prediction and detection of singularities in circadian clocks and in biochemical oscillations. Without going into too much detail, we must first look at the main features of cardiac physiology, the features that bear most on questions of timing.

An adult human heart (Figure 2.1) weighs about one-third of a kilogram. Hydrodynamically it is four-chambered but electrically (our main concern) it is only two-chambered. Each chamber is largely composed of muscle called myocardium ("muscle of the heart"). The myocardium of the two main pumping chambers, the ventricles, is much thicker than the atrial myocardium; in fact the thicker left ventricular wall provides 2 out of the 2½ watts of power used in this pump.

The most frequent spontaneous firings come from the sinoatrial node (Figure 2.1). It initiates each heartbeat in the atria, the "priming" chambers that assist filling of the ventricles prior to their more forceful contraction. The sinoatrial node is the functionally dominant pacemaker for the whole heart under normal conditions. Tissues with a slower intrinsic rate (latent or subsidiary pacemakers) are prematurely triggered by the rapid-fire impulses that normally propagate from that dominant pacemaker. If the dominant pacemaker should be somehow suppressed, then one of the faster among the remaining latent pacemakers (for example, the atrioventricular node, AVN, Figure 2.1) may usurp dominant pacemaker function. Under normal conditions the AV node—the sole electrical junction box connecting atria to ventricles—fires only on command from the SA node, imposing a delay of a couple tenths of a second after

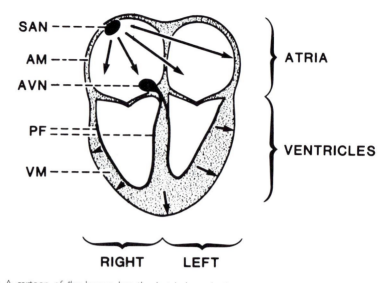

Figure 2.1: A cartoon of the human heart's electrical conduction system, omitting efferent and afferent nerves and the plumbing connections. In order of normal activation: sinoatrial node (SAN), atrial myocardium (AM), atrioventricular node (AVN), Purkinje fibers (PF), and ventricular myocardium (VM).

the contraction of the atria before the ventricles contract. The ventricles are activated from the AV node through especially efficient conducting fibers similar to nerves, the Purkinje fibers (PF, Figure 2.1). They ramify across the inner surface of the muscular wall of each ventricle, triggering it to contract by an electrochemical disturbance equivalent to a nerve impulse (see Figure 2.2, below). There is a so-called refractory interval of one or two tenths of a second (depending on heart rate) after each such triggering. During this time normal stimuli cannot trigger the muscle to propagate an impulse or to contract again, so the impulse cannot backfire. The mass of ventricular muscle normally contracts almost simultaneously, being almost simultaneously electrified all over its inner surface by the Purkinje fibers. Activation is complete as soon as the impulse propagates to the outer surface: usually in less than a tenth of a second. In most areas of the ventricular muscle, conduction of impulses any great distance parallel to the surface does not normally occur, but it prominently distinguishes a great many arrhythmias.

What is the physical basis for this "triggering" and "propagation"? The cells of the heart, like ordinary nerve cells, are electrically polarized, the inside being about a tenth of a volt more negative than the outside. This seemingly trivial potential difference (alias voltage) is very nearly the maximum sustainable without dielectric breakdown (sparking). It amounts to some ten million volts per meter of potential change across

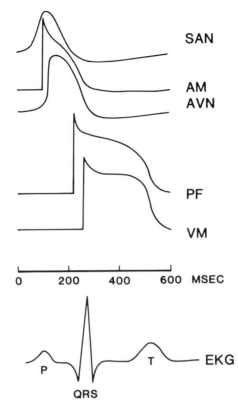

Figure 2.2: Typical traces of membrane voltage against time (in milliseconds, 0 to 600) in five tissues of the normal heart, showing the sequence of activation. Below the time scale is a typical electrocardiographic signal from the chest wall, labeling the letter-named main deflections that correspond to surface gradients in conspicuous events above.

the thin insulating membrane of each cell. The insulation is punctuated by molecular channels that limit the traffic of charged ions in and out, thus maintaining the polarization. An electrical field of that intensity is keenly felt by electrically charged molecules in the membrane; if it should change substantially, they will change shape. As it happens, depolarization beyond a certain limit results in changes that allow an ion current to pass, furthering the depolarization. This process, together with the recovery that follows, is called excitability; the event is called an action potential. Action potentials of several kinds of cell in the healthy heart are sketched in Figure 2.2. Each lasts a few tenths of a second. Some can now be calculated from the underlying equations of ion flow and voltage change. The equations look like Figure 2.3.

The first equation says that the local electric potential, V, changes with

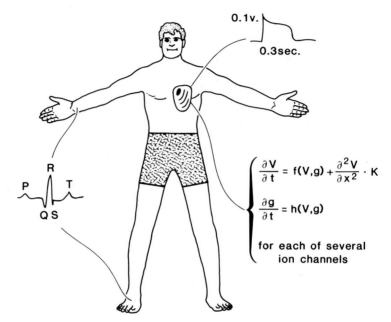

Figure 2.3: Figures 2.1 and 2.2 are put in context, alongside a summary of the main features of the quantitative interrelations among voltage, conductivities, and their rates of change (the equations of electrophysiology). Membrane voltage V changes at a rate proportional to the sum of local ionic currents (f, depending on membrane voltage and local conductivity, g) plus currents from nearby (a physical factor K times the second spatial derivative of voltage). Channel conductivities (g) change at rate depending (h) on membrane voltage and present conductivity.

time ($\partial/\partial t$) at a rate (=) composed of two terms. One term concerns the geometric arrangement of nearby potentials ($\partial^2 V/\partial x^2$); it computes the electrotonic currents that couple membranes across space. Its coefficient, K, depends in a simple way on the passive electrical conduction properties of the cell; it has the same dimensions (cm²/sec) as a chemical diffusion coefficient. This will become important in Chapter 7. The other term adds a local ionic current (f) that depends on the local potential (V) and the local openness (g) of each ion channel.

The several other equations are simpler: they describe the rates of change ($\partial/\partial t$) of openness (g) of each ion channel as a mathematical function of present openness and local potential. These functions are slightly different from one cell type to the next, accounting in part for the different shapes seen in Figure 2.2. In particular, some cells will slowly depolarize and then smoothly fire spontaneously, as can be seen, for example, in the upper trace: sinoatrial node (SAN). The Purkinje fibers (middle trace) would too, given a chance, but their spontaneous firing is normally anteceded by arrival of triggering action potentials from the

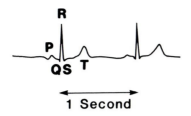

1 Second

Figure 2.4: Two cycles from one of the twelve electrical pickup leads of a normal electrocardiogram (precordial lead V6, 60 beats per minute, the author's).

atrioventricular node (AVN). This precocious triggering, seen in the abrupt vertical segment of the PF trace, for example, finishes the accounting for the different shapes in Figure 2.2. PF, unlike the AVN, which is also triggered but shows a less abrupt rise, commence to fire with a rapidly regenerative inrush of sodium ions, hence the abruptness of depolarization (upstroke) in the middle trace. Like the atrial muscle fibers (AM), most ventricular muscle fibers (VM) might never fire spontaneously: they need a little nudge from the Purkinje fibers (bottom trace).

Implicit in these equations are the excitability, the rhythm, the changes in timing imposed by a momentary partial depolarization (nominally an excitatory stimulus) or hyperpolarization (nominally an inhibitory stimulus), and the spread of action potentials across the heart. That moving depolarization induces faint currents throughout one's saline insides. It can be detected outside the skin, just as the mechanical correlates of the heartbeat send sound waves that can be detected outside the skin. Doctors monitor both, by stethoscope and by electrocardiograph, to check the regularity of the heartbeat. Figure 2.4 shows a sample from the author's electrocardiogram (EKG). Its particular appearance depends, of course, on the placement of pickup electrodes on the skin, but the same deflections appear at the same intervals in one guise or another in any case. In this particular arrangement the main named deflections are conspicuous. The P wave reflects depolarization in the thin-walled atria. The QRS wave reflects depolarizations of the thick-walled ventricles slightly later. The T wave reflects repolarization of the massive ventricles. (Atrial repolarization is too faint to see and, anyway, occurs just about when the ventricles fire.) More exactly, it reflects a brief (1/10 sec) accentuation of spatial voltage gradients during repolarization; were recovery uniform, this electrical artifact would be self-canceling. Everything to be analyzed henceforth depends crucially on the patterns of spatially graded timing and of stimulus impact. In Chapter 3 we will linger over the significance of this "vulnerable phase" marked electrically by the T wave.

Then there is a quiet time—an interval of flatness between beats when

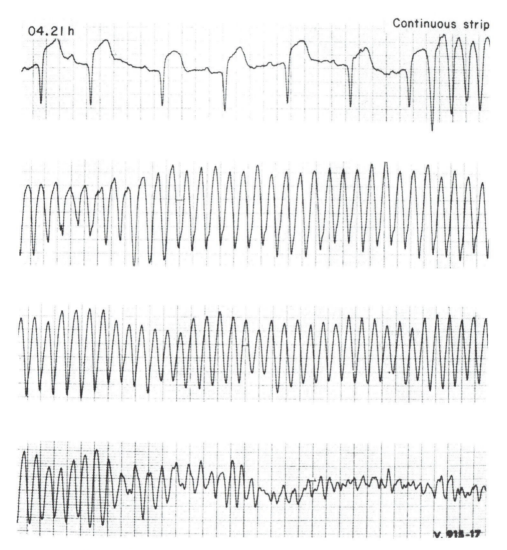

Figure 2.5: Electrocardiogram during sudden cardiac death, with faint grid markings 1/5 second apart (from Roelandt et al. 1984). The patient was known to be at risk and was wearing an ambulatory EKG recorder at the time of his misfortune. Normal sinus rhythm at roughly 1-sec intervals was interrupted by an early premature ventricular contraction (PVC, protruding from the bottom right of the uppermost segment of the trace) during the vulnerable phase (see Chapter 3). Tachycardia ensues at about 5 Hz, soon degenerating into terminal electrical activity that looks irregular when summed into this electrical trace.

essentially nothing is going on electrically, but the atria are filling—until the sinoatrial node sets off the next P wave. This at least, is the normal situation. In contrast, in the presence of a phase singularity, as we shall see, the equations of excitability describe a perpetually circulating wave that allows no rest (and pumps no blood). In this case, from the moment when serious trouble starts, throughout the dysrhythmic prelude to fibrillation, until the heart finally stops beating for want of oxygen, there are no more flat parts in the EKG: a wave of depolarization is always somewhere present on the heart. Figure 2.5 shows such a trace automatically recorded by a portable "Holter monitor" in one published case of sudden death.

What structures are at fault in the initiation of such an event? All parts of the heart, but especially the pacemaker nodes and the specialized conducting tissues, are infiltrated by a great variety of nerves conveying impulses from diverse internal organs indirectly through the central switchboard, the brain. The vagus nerve, for example, is a pair of cables from the brain containing fibers of many kinds that arborize in the atria and in the specialized conducting fibers of the ventricle. These impulses have diverse effects on the local rates and mutual coordination of the many parts of the heart. They shorten the refractory period slightly and accentuate spatial gradients of repolarization. The nonuniformity of sympathetic innervation of the ventricles, likewise, results in irregular gradations of automaticity, of diastolic threshold, and of refractory period during times of sympathetic activity (stress) [Bellet 1971; Randall 1977; Kralios et al. 1975]. The myocardium is of course also infiltrated by the oxygen-carrying capillary beds of the coronary circulation; an abrupt change in their rate of perfusing the muscle might also both provide a more gradual stimulus and create additional gradients of timing.

All this machinery commonly functions in a well-synchronized regular rhythm. When damaged, for example by an obstruction in its blood supply (coronary occlusion), the rhythm can fail, the deprived patch of muscle turning ischemic and inexcitable while fibers in the fringe areas are abnormally active. Even when undamaged, the heart's normal synchrony can also fail during exposure to unusual levels of normal hormones and ions. Even without such stress, it can become turbulently disorganized when jarred by unusual stimuli, as in accidental electrocution or a local failure of blood supply, or even by usual stimuli, timed unfortunately. The common causes of transition from normal synchrony to turbulent disorganization conspicuously include stimuli that intrude upon or even cause a sharp local phase dispersion among formerly synchronous fibers. Such dispersion, not random but characteristically organized in space, is the essence of a phase singularity.

Fibrillation

J. A. MacWilliam coined the term *sudden cardiac death* a century ago [MacWilliam 1888].[1] The phenomenon was not new. Such untimely departures have been noted and deplored for many centuries. But it was not until the nineteenth century that MacWilliam argued that the heart did not just stop in sudden cardiac death: something first triggered it into a futile apoplexy: "the cardiac pump is thrown out of gear, and the last of its vital energy is dissipated in a violent and prolonged turmoil of fruitless activity in the ventricular wall" [MacWilliam 1888]. This condition in the atria, rather than the ventricles, is clinically observable because it does not cause instant death; it was then called delirium cordis (madness of the heart); its modern name, fibrillation, was originally a French term invented in 1874 by Alfred Vulpian, connoting his impression that the individual fibers were contracting independently. To me the hot, wet heart of a dog held in my hand resembles, when fibrillating, nothing so much as a frightened bird unable to flap its wings but quivering and shaking uncoordinatedly from top to bottom. The heart fibrillates "like a wad of writhing worms. In many cases the rate of this random process is so rapid that the heart surface seems to shimmer. In other instances multiple waves of contraction and relaxation are clearly visible" [Tacker and Geddes 1980].

There are many modes of fibrillation; the term is generic, much like "turbulence" in fluid mechanics. As in fluid mechanics there is much debate about what really qualifies as turbulence or fibrillation, according to diverse test criteria, and there are increasingly complicated stages as prefibrillatory arrhythmias (vortex modes) progress to full-blown fibrillation (turbulence) [Swinney 1985]. In advanced stages of fibrillation, microelectrodes show little local synchrony except at points 1 mm or less apart [Bellet 1971; Herbschleb et al. 1983], possibly because at such stages circulation has long failed and the junctions that normally couple cells electrically are also beginning to fail. In this book we will be more concerned with the onset and earliest stages, comparable to simple vortex instability in a fluid, when "fibrillation" is scarcely distinguishable from "paroxysmal tachycardia." (Tachycardia means "fast heartbeat," technically at a rate exceeding 100 beats per minute [bpm] but, as used here, usually exceeding 300.)

Fibrillation is easy to start. It is easier in muscle that is already relatively inexcitable ("depressed"), as when deprived of normal blood flow or

---

[1] The term today is sometimes used broadly, as in this book, and sometimes [e.g. in Kannel and Thomas 1982] in a deliberately restrictive way to encompass only those cases (the majority by far) in which the victim was already at risk due to badly obstructed circulation.

driven to contract at an abnormally high rate. But it is easy enough in perfectly normal muscle. The most convenient artificial means is electric current from a reading lamp: the 50–60 Hz we chose for our electric standard is nearly the "optimum" frequency. A momentary touch on the exposed surface of a dog's heart, even with the voltage turned way down, instantly transforms the former rhythmic contractions to uncoordinated shivering—and not only at the spot touched but all over the heart, quicker than you can see it spread. There is still no widespread agreement on the mechanism of fibrillation, perhaps because there are many varieties: due to high-frequency excitation from multiple foci, due to excitation propagating irregularly in the meshwork of Purkinje fibers lining the ventricular cavities, due to vortex-like excitation in the three-dimensional ventricular muscle mass, due to any number of other postulated mechanisms. Disagreement on mechanism may also stem from the difficulty of description, as with turbulence in water. Like turbulence, fibrillation matures through a succession of stages: during the course of a few seconds it progresses from a gross tachycardia to coarse rapid undulations, often markedly periodic and responsive to countershock, to the still quicker fine-grained shimmering described above that is often refractory to purgative measures. It seems to get started by some kind of electrical instability: at least in healthy muscle, it is an organizational problem more concerned with the electrically coordinated relative timing of activity in the heart's numerous fibers than with any initial defect in their individual behaviors. Except for the finding that the fast sodium gates cease to function in fibers triggered too frequently, disappointingly little has been learned about fibrillation itself from electrophysiological studies on normal single fibers. However much can be inferred (see below) about the prelude (tachycardia) and the diverse mechanisms of its triggering. The essential feature in common among all mechanisms is that fibrillation is caused by factors that throw fibers out of phase with one another. Part II will dwell upon that fact and the detailed spatial pattern of those phase discrepancies.

Whatever the immediate causes of its onset, fibrillation is the commonest portal to sudden cardiac death. In most instances, factors that opened the portal are obvious to the coroner. Acute blockage of a coronary artery is especially common; in such cases, the affected fibers are decidedly not functioning normally, if at all. In other instances—cases of electric shock or emotional shock among them—the triggering malfunction may have been tiny and transient, leaving no visible trace. No clue lingers beyond the heart's last motions: the coroner confronts a body cold and blue, often with undiagnosed atherosclerosis that may have made the heart electrically unstable, but finds nothing else amiss. What happened to trigger the instability? Why are healthy people suddenly

Figure 2.6: G. R. Mines in his McGill University laboratory, 1913. Courtesy of David A. Rytand, M.D., Stanford University Medical Center.

———— TIME ————▶

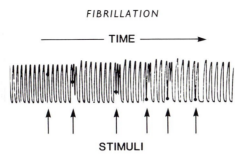

STIMULI

Figure 2.7: Transient derangements of rabbit heartbeat following an electric stimulus to the apex of the ventricle at various moments in the cycle. From Mines 1914.

staggered by abrupt disorganization of their heartbeat? How can the parts of a human heart so quickly lose their lifelong mutual synchronization? If this anarchy persists for as long as five minutes, leaving the unaided victim no means to nourish his brain with oxygen, the lapse of consciousness becomes irreversible. (Of course bystanders need not stand by; CPR technique can keep the brain oxygenated until more medical aid arrives.) What was the triggering stimulus? For the present we have before us only a diversity of tantalizing clues. Anyone who shares the curiosity of children and scientists cannot escape the impression that we have here the pieces of a single jigsaw puzzle.

Their assembly was almost in hand as early as 1914 [Mines 1914]. G. R. Mines (Figure 2.6) was widely considered the most promising young physiologist of his generation. At the time of his death, under circumstances mysterious enough to intrigue Sherlock Holmes,[2] Mines had great hope of penetrating the mystery of fibrillation. He had tried to upset the usual course of a rabbit's heartbeat by delivering a single electrical stimulus with a tap of a telegraph key. He probed the entire cardiac cycle from one beat to the next by giving a stimulus, watching the aftermath until normal rhythmicity resumed, then giving another, timing each successive stimulus a little later after the preceding beat. Most such stimuli had no lasting effect: the little rabbit heart was deranged in its beating for less than a second, then returned to its former regular rhythm, as expected.

The contractions of the rabbit's heart were conveyed by a fine thread to a lever. At the lever's opposite end, a recording surface moved steadily under a pen point. Figure 2.7 shows a few segments of Mines's original recordings—the permanent tracing of the pen point's lurching up-and-down movements as the thread pulled and relaxed in rhythmic repetition.

[2] Obituary reports that Mines was recording his own heartbeat at the time of his death in his laboratory led to the speculation that he might have been experimentally perturbing his heartbeat [Winfree 1983]. Information obtained recently from the McGill University Physiology Department makes this seem less likely.

The regular rhythm was interrupted from time to time, starting with a tiny black dot. Each of these dots marks an electrical spark conveyed by the pen point to the recording surface at the moment when Mines closed a telegraph key; at that same instant a milder electric stimulus struck the apex of the beating heart.

This untimely artificial stimulus interfered with the usual sequence of neuroelectric events coordinating the heartbeat: the lever movements are momentarily deranged. But of all organs, the heart is best evolved for persistent regularity in the face of every kind of disturbance, and it swiftly recovers its composure: the regular beat almost always returns. Almost always, but not always. Mines may have been the first to discover a peculiarity of the human heart that is responsible for many tens of thousands of tragically unexpected deaths each year. Many victims of sudden cardiac death might be living today had he survived to understand that discovery: that a fleeting shock can sting even a healthy human heart into fatal turbulence. In Mines's pioneering experiments there lurk two clues to the nature of fibrillation, one involving time (Chapter 3) and one involving space (Chapter 5). We will examine the ragged edges of these two clues and see if they fit together. The result illuminates one process by which the soil is prepared and the seeds of fibrillation—here interpreted as phase singularities—may be accidentally planted.

A Clue Involving Time

I have no data yet. It is a capital mistake to theorize
before one has data. Insensibly one begins to twist
facts to suit theories, instead of theories to suit facts.
—Sherlock Holmes, in
"A Scandal in Bohemia"

The first of Mines's two clues turns on the words *almost always*. The
heart beat almost always returned; in other words, sometimes, rarely, it
failed to return. Figure 3.1 shows two more tracings from his posthumously published paper. Notice the abrupt termination of coherent contraction following the right stimulus given at just the right time. This is
fibrillation. Every fiber of the heart is still beating, in fact severalfold
more often than before. But the myriad muscular fibers have lost their
simple coordination. They no longer tug in unison at the pen. They no
longer beat synchronously, rhythmically, but rather in a different pattern
so fine-grained and so rapid that the pen only trembles. Mines found this
reaction to his stimulus only when it was big enough and when it was
timed exquisitely: notice that the black dots that lead to fibrillation appear
just as the heart is relaxing after a spontaneous contraction, before the
next begins. According to modern measurements [El-Sherif et al. 1977]
the timing must be good to within 30 milliseconds, in the electrocardiogram's T wave. As Mines [1914] wrote, "under some conditions a
stimulus of very brief duration may induce fibrillation. . . . a single tap
of the Morse key *if properly timed* would start fibrillation. . . . the stimulus employed would never cause fibrillation unless it was set at a certain
critical instant."

The Vulnerable Phase

This narrow window in time between two successive beats plays a dominant role in medical research papers today about sudden cardiac death.

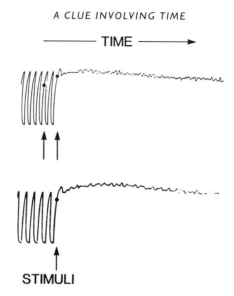

Figure 3.1: As in Figure 2.7, but fibrillation ensues when stimulus of sufficient size (but not too strong) is administered at a critical moment that later came to be called "the vulnerable phase."

When rediscovered decades later, this special time between beats when a single stimulus can reduce the future to chaos became known as the vulnerable phase.

But Mines only discovered that the heart has this Achilles' heel, a critical moment when it is susceptible to disorganization by a stimulus of the right size. He did not resolve anything about the nature and origin of this chink in the armor. What kind of spear must whistle through this window of vulnerability to have such a startling effect? What kind of stimulus does it take to shatter the normal organization of timing at the one moment when the heart is susceptible? Mines used electric shock, but not to study accidental electrocution; he was using simple electrical equipment to *simulate* the neural stimuli that regulate the pace or force of our heartbeats throughout the day as our workloads and excitements vary. Any mammal's heart is always subject to untimely stimuli along the pathways of the autonomic nervous system from the brain and its nervous input tracts [Richter 1957; Burch and DePasquale 1965; Armour et al. 1972; Hageman et al. 1973; Kralios et al. 1975; Lown and Verrier 1976; Lown et al. 1977; Corr and Gillis 1978; Randall et al. 1978; Engel 1978; Lown 1979a; Lown 1979b; DeSilva 1982; Korner 1979; Verrier and Hagestad 1985]. Even a temporary and local obstruction of coronary circulation might provide an electrical perturbation by allowing the ionic balance of affected cells to run down. In small mammals the remote chance that any such stimulus might trigger fibrillation is only a remote chance of fainting (syncope): little hearts spontaneously recover their synchrony after a while, as you see in Mines's recording of the tiny rabbit heart. But a heart as big as man's is big enough to sustain widespread turbulence. The turbulence can be perniciously persistent, once established. According to several estimates, ventricular fibrillation claims four to sixty lives hourly in the United States alone. Many such events represent lethal encounters with the ventricular vulnerable phase [Surawicz and Zumino 1966; Hinkle et al. 1977; Adgey et al. 1982; Hohnloser et al. 1984; Geuze and Koster 1984; Roelandt et al. 1984; Figure 2.5, above]. (There are of course many other means of inducing tachycardias and fibrillation that apparently do not implicate the vulnerable phase. There is also an atrial vulnerable phase, but one can still live with fibrillating atria.)

Why does such a moment even exist? If it only amounts to 30 milliseconds, why hasn't natural selection somehow eliminated the problem? Could there be some connection with the topological phenomena we explored in Chapter 1, the unremovable phase singularity and the critical stimulus that conjures it?

Phase Resetting in Electrical Pacemakers

The patterns of phase resetting discovered in Chapter 1 are not unique to circadian rhythms. They are the patterns typical of a broad class of oscillator mechanisms called attracting limit cycles (see Appendix), and of an even broader class in which populations of loosely coupled oscillators generate a collective rhythm [Winfree 1975a; Winfree 1976; Winfree 1980]. The electrical oscillations of pacemaker membranes belong to the same mathematical class, even though the molecular mechanisms involved are surely quite different from those underlying circadian clocks. Thus there seems to be a reason to look here for a familiar pattern.

The easiest way to recognize that pattern is first to recognize the even mode of phase resetting. Do neural and cardiac pacemakers react to neuroelectric stimuli in that topologically distinctive way? Up to the mid-1970s the question had never been asked, though in the case of neural pacemakers the data had been collected and published with other questions in mind and only awaited reinterpretation.

Even Resetting in a Pacemaker Neuron

Just as in the circadian rhythms literature, even resetting was first discovered a decade before it was recognized. The discoverers were Theodore Bullock, Donald Perkel, Joseph Schulman, Jose Segundo, and George Moore in Los Angeles. The subject was a spontaneously rhythmic neuron in the crayfish. This particular pacemaker, called the stretch receptor, is

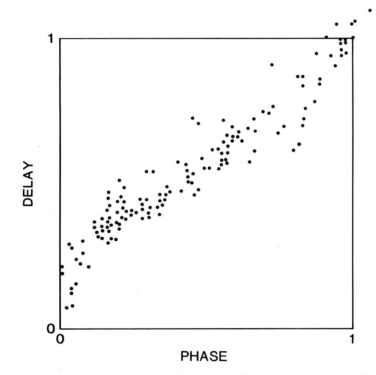

Figure 3.2: A hyperpolarizing neural stimulus to the crayfish stretch receptor delays its next firing by an amount that increases the later the stimulus is given after prior action potential. Adapted from Perkel et al. 1964 and from Winfree 1981 by permission of the American Mathematical Society.

a 1-cell strain gauge on a muscle. It reports muscle tension as a period between firings (normalized to 1 in Figure 3.2). It also receives inhibitory input from an attached nerve fiber. Its next firing is delayed when an action potential arrives along that fiber to provide an inhibitory (hyper-polarizing) stimulus; the amount of delay depends on the timing. If the delay of the next firing is plotted against input timing, there is a big discontinuity near the end of the cycle because input later than that moment cannot prevent the regularly scheduled next firing, and then the same pattern repeats from the beginning of the next cycle. Phase 0 in Figure 3.2 represents the moment of pacemaker firing. Phase 1/2 is mid-way between two undisturbed firings. Phase 1 is the moment of the next firing, the same as phase 0 of the next cycle. The data points show that the longer the wait after a spontaneous firing before the next stimulus is given, the greater the delay of the next firing.

Viewed this way, it might appear that any curve sketched through the data cloud must sharply descend at phase $1 = 0$ on the right to reach small delays such as we observed at phase 0 on the left. Such a steep fulfillment of the dictum that what goes up must come down would not seem surprising, since the nerve membrane does undergo an abrupt rearrangement during its action potential. Moreover, such a curve would represent the familiar kind of resetting (odd, weak) that everyone expects to see if the curve is continuous at all, at least in the limit of small stimuli. We will see in a moment that this expectation is not fulfilled: next-firing time changes discontinuously as old phase increases, and the new phase of the rhythm is reset in a smooth and fundamentally different way.

In Chapter 1 a graphical language was used to describe the changes of timing. In this chapter the same ideas and diagrams inevitably recur, but the language is revised somewhat to conform to the usage of electrophysiologists. "Stimulus" means any disturbance of the normal membrane process, whether it be inhibitory, subthreshold excitatory, actually elicits immediate firing, or has more subtle effects. The time from the previous beat to the stimulus is what neurobiologists call the coupling interval. Measured in units of one normal cycle, it is also often called the "old phase." The time from the stimulus to the next beat is called the first latency. The new phase is the natural period minus the latency, again, measured in units of one normal cycle duration (Figure 3.3). There is, of course, also a "second latency," measured from the stimulus to the second subsequent beat, and a third latency, and so on. Any latency plus the new phase is an integer number of periods. The "cophase" of a reset circadian rhythm is the latency, minus any complete periods contained within it, measured in units of one period; cophase plus new phase = 1.

The format of Figure 3.2 leads us to look for a descent at the end of the cycle, but where is it? There is no descent. Moreover, when the data are replotted as in Figure 3.4 (plotting the directly observed latency upward or new phase downward, rather than subtracting expected latency from observed latency to get "delay" before plotting), their continuity in the even pattern is evident. The middle horizontal line indicates latency 0, the time of the stimulus. The diagonals with dots on them below time zero indicate pre-stimulus firing times, and their dotless extensions above time zero extrapolate that pattern to anticipate next firings, had the stimulator accidentally been left unplugged. The delayed dots at the top of the figure are the data of Figure 3.2, transposed, then duplicated to the right to fill most of a second cycle of coupling intervals.

Notice that just as in Figure 3.2, as coupling interval increases to the right, the data dots fall later and later after the anticipated firing time (greater delays), reaching full cycle delay at coupling interval = 1. Then they continue smoothly. There is no discontinuity, no steep decrease in amount of delay near phase 0. Instead we see only a smooth connecting rise. The response that was termed a big delay changes as it must early

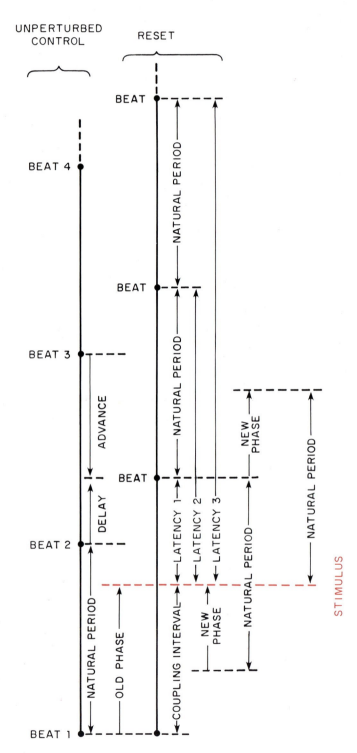

Figure 3.3: Along parallel time axes, we see idealized perfectly periodic action potentials ("beats") in an unperturbed control experiment and in a phase-reset replicate. Beats in the perturbed rhythm are left unlabeled (no "1, 2, 3") to avoid any implication of unique correspondence with beats 1, 2, 3 of the unperturbed control. For definitions of new phase, old phase, and latency see text and Glossary.

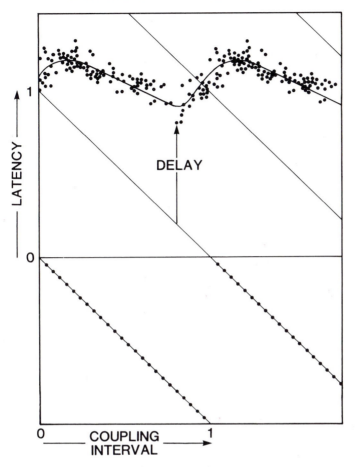

Figure 3.4: The data of Figure 3.2 are reconstructed to infer the original experiment, prior to subtracting control from reset to obtain "delay." Result: latency vs. coupling interval looks smooth. The stimulus is given along the bottom horizontal line at times later and later after the prior action potential in (otherwise) replicate experiments aligned side by side from left to right. Data are reduplicated beyond the end of the cycle merely to show how the end connects to the beginning without discontinuity. This connection is redone in Figure 3.5. Adapted from Winfree 1981 by permission of the American Mathematical Society.

in the cycle to a small delay, not because there is any drastic change in firing time, but merely because when we passed the moment of a second anticipated firing, we implicitly switched to using *that* firing as the new reference time for evaluating the amount of "delay."

Viewed this way, the interpretation of advances and delays seems to tell less about the underlying timing than about the naming of action potentials: when the "first" is no longer suppressed, that name pops back, leaving the later firing now to be called "second."[1] This is the familiar international date line paradox: fish swimming under the line may pass from 3:00 P.M. Tuesday to 3:00 P.M. Wednesday, but they won't feel any change because the ostensible discontinuity of time is only due to a naming convention. If you look at the latencies of later events, ignoring whether the event is called "fifth" or "fourth," you see that it varies smoothly with the coupling interval and it varies in the even pattern. (This is often called "strong" resetting in the literature of neurobiology.)

To plot one periodic quantity (a phase) against another, we want graph paper that is periodic in both directions. The surface of a bagel, doughnut, or inner tube—generically, a "torus"—serves admirably. Just as with circadian clock resetting [Winfree 1980], the neural data could be plotted on a torus of "new phase" vs. "old phase" or of coupling interval (deleting complete periods, if any) vs. latency (again, deleting complete periods, if any). In that format, control data (arbitrarily faint stimuli; new phase = old phase; coupling interval + latency = unit period) lie on a closed ring linking through the hole in the torus. In contrast, the resetting data of Figure 3.2 lie on a smooth equatorial ring that could be lifted free of the torus, because it distinctly does not link through the hole: as old phase runs full cycle (coupling interval moves from 0 to 1), new phase varies little (because latency varies little) (Figure 3.5). This topologically distinctive feature—later discovered in many other pacemaker preparations—was not commented upon until thirteen years after the data first appeared without remark in the same journal [Perkel et al. 1964; Winfree 1977].

Resetting the Heartbeat

What about the electrically rhythmic membranes of the heart? Are they so different as to never exhibit even resetting? The role of phase resetting in the timing of cardiac pacemakers was first noted by Donders [1868] but was never systematically pursued until the 1930s, by 1963 Nobel Laureate Sir John Eccles. Like Mines, Eccles was examining the effect of

---

[1] If your watch is 8 hours wrong (perhaps because you have just crossed 8 time zones) it is not useful or even meaningful to distinguish whether it is "advanced 8 hours" or "delayed 16 hours." However, in some situations the transient process of resetting may assume dominant importance, e.g. when an oscillator is unable to recover to its normal rhythmic behavior between one perturbation and the next during periodic stimulation intended to entrain. In such cases information is lost by the "modulo" operation implicit in restricting attention to the rhythmic phase aspect of timing. Throughout this book, however, we deal exclusively with a single discrete perturbation—of arbitrary duration and complexity, but applied only once. And the oscillators we deal with do recover almost back to their usual cycles before we will need to assay their resetting.

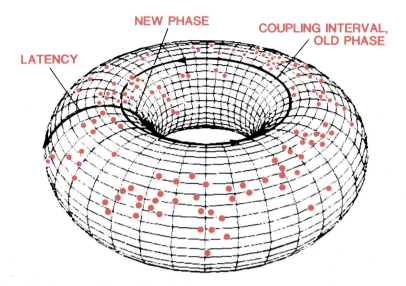

LATENCY

NEW PHASE

COUPLING INTERVAL, OLD PHASE

Figure 3.5: The data of Figures 3.2 and 3.4 are replotted on toroidal coordinates to emphasize their peculiar topological connectivity. Action potential time is taken as 0 in both coordinates from the intersection of axes on the top left (the origin). Coupling interval or old phase increases counterclockwise from there in the horizontal plane, while new phase increases clockwise from there in the vertical plane (latency therefore increases counterclockwise). The new phase = old phase (or coupling interval + latency = 1) control data would fall along a diagonal ring through the origin linking this torus; almost all of it is hidden in this projection. The experimental data, in contrast, are all visible, lying in a ring along the top of the torus.

stimulus timing on the subsequent heartbeats. Unlike Mines, he was not looking for fibrillation but only for changes in the timing of the next few beats. And his stimulus was not an electrical shock to the apex of the ventricle but something more natural: activation of the dog's vagus nerve, one of the normal regulators of the rate of beating. Eccles and his collaborators G. L. Brown and H. E. Hoff at the University of Oxford observed that the pacemaker was at first disturbed after a stimulus but soon returned to its usual regularity, much as in most of Mines's recordings [Eccles and Hoff 1934; Brown and Eccles 1934a; Brown and Eccles 1934b]. In many cases the pacemaker recovered so swiftly that every complete cycle after the stimulus was of normal length: the principal lasting effect was only an offset of timing, for example, a delay by a third of a cycle. The rabbit heart used by Mines is even more consistent in its cycle length.

In the case of a negligible stimulus, the next beat is of course not at all displaced, so the coupling interval plus the latency (the old phase plus one period minus the new phase, but new phase = old phase in this case) equals one normal cycle length. As we vary the coupling interval of this negligible stimulus, increasing through one full cycle, the complementary latency must (and does) smoothly decrease through one full cycle, as shown in the reference diagonals of Figure 3.4. Every possible value of latency is found an odd number of times per cycle (namely, once in this case). This case merely illustrates the format of data plotting; it is a "control" experiment, showing what would happen in the normal case if all the experimental arrangements were in place, but someone forgot to plug in the stimulator.

The Two Distinct Styles of Phase Resetting Revisited

A real experiment is done with the stimulator plugged in, of course. What happens? Things go much as before but the non-zero stimulus noticeably alters the timing of the next beat. In Eccles's early experiments, that timing changed discontinuously as the coupling interval was varied. In the intervening half-century scores of neural and cardiac pacemakers have been found to behave this way.

Others, however, show apparently continuous variation in a particular pattern that arouses deep curiosity: the smooth "even resetting" pattern. Such results are beautifully illustrated in an experiment by José Jalife and collaborators at the State University of New York Upstate Medical Center [Jalife et al. 1983]. Here the timing of a rabbit's sinoatrial pacemaker is perturbed by an electrical stimulus that triggers the nearby vagal nerve endings to release acetylcholine. As in the crayfish stretch receptor, this stimulus is hyperpolarizing (inhibitory); but unlike the crayfish preparation, in this one the magnitude of the stimulus can be conveniently varied by varying the duration. Timing is delayed variably as the coupling interval varies, but still the latency decreases through one cycle while the coupling interval increases through one cycle (Figure 3.6). Each duration of latency (neglecting multiples of a full cycle) is represented an odd number of times, as in the control experiment with no stimulus at all. In keeping with the notation of Chapter 1 (and to contrast it with what comes next) this pattern is called odd resetting.

Next comes a seemingly innocuous but rather curious fact. If the stimulus is big enough (but still in the physiologically realistic range of 50–100 msec vagal bursts), the latency varies *less* than one cycle: it only increases and decreases back to its original value as the coupling interval increases through a cycle (Figure 3.7). In this case the latency varies through *no* full cycles, and only some values occur, each an even number of times (namely, twice or not at all in the data illustrated). In fact, this latency vs. coupling interval plot superimposes almost exactly on the one obtained twenty years earlier in the crayfish stretch receptor's response to inhibitory synaptic input. This pattern, qualitatively different from

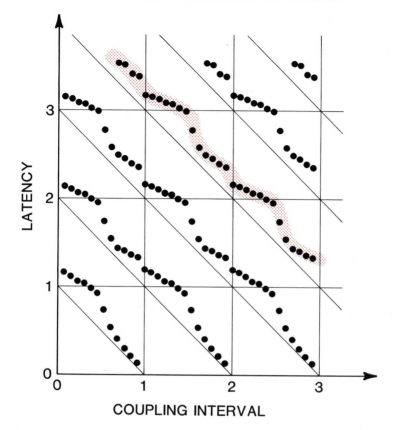

Figure 3.6: In the format of Figure 3.4, data from Jalife et al. 1983 are triple-plotted to the right to show odd resetting of the sinoatrial node by vagal stimulation. Each vertical trail of dots is one experiment. The diagonals are extrapolations from pre-stimulus firing times at the standard period of about 400 msec: data would fall on these lines were the stimulus ineffective. As it is, there are slight delays, varying somewhat with the coupling interval of the stimulus. Though the resetting curve (red stipple) is continuous, it pieces together segments of first latency with second with third and so on. This figure, 3.7, and 3.8 are adapted from Winfree 1983 with permission from *Scientific American*.

Figure 3.6, is called (just as when we found it in Chapter 1 and in Figures 3.4 and 3.5) even resetting. To underscore its distinctness, an extreme case might be imagined in which latency stays the same: the next beat occurs one second after a big stimulus, regardless of when it is given.

Note the little trick lurking here: in both Figures 3.6 and 3.7 (unlike Figures 3.4 and 3.5), it is not exactly the first latency that varies smoothly in the even resetting pattern, or the second, but a curve that smoothly joins them. If you fix attention on the first latency, or the second, both the even and the odd resetting patterns seem discontinuous (Figures

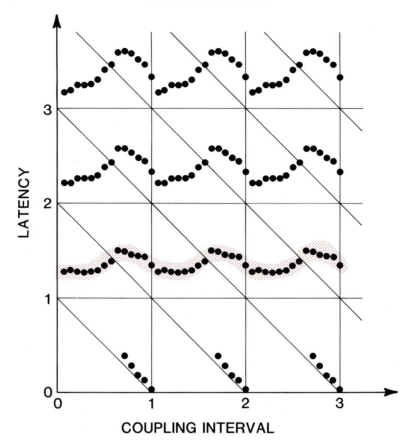

Figure 3.7: Exactly as in Figure 3.4, but the stimulus is slightly greater. Resetting changed to even mode. Note that these horizontally periodic curves (red stipple) cut any horizontal line an even number of times and cannot be bent into the diagonally periodic curves of Figure 3.6, cutting any horizontal line an odd number of times, by any continuous process. Note also that the curves are slowly changing with time after the stimulus: depending on when the stimulus arrived, it slightly increased or decreased the period of the pacemaker, in addition to phase-shifting it. This effect, presumably due to the acetylcholine channels remaining open for a long time (about twice the pacemaker period) after the nominal "stimulus" ends, distorts the resetting curve but cannot change its topology. See Figure 3.6 for source.

3.7 and 3.8) in the same way that misled circadian physiologists until the 1960s. But the data are not discontinuous; only their names ("first," "second," "advance," "delay") are. Revealingly, the ostensible discontinuity spans close to one cycle; no data points ever appear in the full-cycle gap.

It seems there is no steep reversal of a curve or a discontinuity of reset phase here (except in terms of the naming, as above). Rather, the phase data can be connected smoothly in a topologically different pattern that

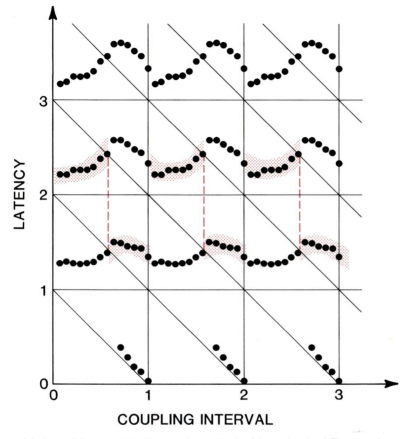

Figure 3.8: Figure 3.7 is repeated with second post-stimulus firings painted red. The smooth even-resetting curve links first and second latencies, second and third, and so on. The resetting curve of second latencies is *not* continuous. See Figure 3.6 for source.

takes no notice of names. On a torus plot like Figure 3.5, the data fall on smooth rings that do (Figure 3.6) and do not (Figure 3.7) link through the hole. We needn't be distracted by the unit jump in enumeration of events. As in crossing an international date line, the renaming occurs when the stimulus falls after a critical stage in the cycle and thus manages to elicit one more spike than it could elicit when applied earlier; but the timing of those spikes is not appreciably altered.

Not all resetting experiments give smooth curves, of course; some do have jumps too steep to distinguish practically from mathematical discontinuities (Box 3.A). But that is another matter. The essential point here is that the smooth results that emerge in even format allow us an extravagant inference.

Box 3.A: Are There Discontinuities in Neural Resetting Curves?

The short, empirical answer is yes. Some are catalogued in Winfree 1980, and more in Glass and Winfree 1984, Guevara 1984, and Guevara et al. 1986, but the evidence usually shows only that the curves are at least very steep in places.

Are any such curves demonstrably discontinuous? This question was first answered experimentally by Michael Guevara in his thesis. Using microspheres composed of many chick heart pacemaker cells, Guevara observed the predicted odd resetting response to weak depolarizing electrical stimuli (less than about 6 nanoamperes in a 114 micron sphere), and even response to larger stimuli (more than about 8 nA, in any case never exceeding a minute fraction of the normal peak sodium current). Within a range of stimulus intensities believed to be intermediate (around 24–27 nA in a 149-micron sphere), the resetting curve appears to have a "jump" in it. To examine this ostensible discontinuity with care, Guevara repeatedly monitored the membrane potential following a stimulus 143 msec after the prior action potential in a 615-msec spontaneous rhythm. An action potential always followed the stimulus immediately, somewhat advancing the rhythm. Applied at 141 msec, the same stimulus always failed to elicit immediate firing, but slightly delayed the rhythm. Does this region hide an incredibly steep segment of a continuous resetting curve? The same stimulus at 142 msec always produced either the one result or the other, but never any intermediate. This looks like discontinuity, without a steep segment.

Guevara's demonstration has particular interest for the exquisite care with which his experiments addressed the question of continuity. The question is significant because it is widely accepted that differential equations in the celebrated Hodgkin-Huxley paradigm do render an excellent description of electrical events in spontaneously active cell membranes. It is also widely assumed that in a continuous dynamical system new phase must vary continuously, even if steeply, with old phase, except at isolated equilibrium states or on a locus of codimension 2. But this is not correct, particularly if "new phase" is assayed from the first latency rather than asymptotically long after the stimulus [Kawato and Suzuki 1978; Kawato 1981]. Many of the differential equation models of electrophysiology have phaseless loci of codimension 1, e.g. a model with three equilibria examined by Clay et al. 1984, Glass and Winfree 1984, and Guevara et al. 1986. Thus a demonstration that for some range of stimulus magnitudes the resetting curve is not only exceedingly steep but actually discontinuous does not in itself impugn the reliability of differential equations of electrophysiology.

Nonetheless Guevara also suggests that the discontinuity might derive ultimately from the discreteness of individual sodium channel events within the vast expanse of membrane affected, and that both differential equations and the continuity ideas of dynamical systems theory (see Appendix) may be inappropriate in this case.

Rigorous efforts like Guevara's are necessary to document discontinuity in the resetting pattern of an oscillator after it has returned to normal rhythmicity. But if one examines only the transient immediate aftermath of perturbation—e.g. the latency to the first action potential after a stimulus—there is no reason to expect continuity. Kawato 1981 first made this distinction clear in mathematical terms.

A Hidden Consequence

In order to see the startling but hidden significance of this resetting style, attention must be focused on the timing of heartbeats relative to the stimulus. As in the analogous experiments using circadian clocks, we want to neglect any complete cycle–durations in the latency time (first, second, nth: all the same but for integer multiples of the standard cycle duration; it makes little sense to distinguish whether my watch is 1 hour slow or 25 hours slow or 23 hours fast). A tidy way to strip out the superfluous full periods is to take the interval of latency and wrap it around a ring of circumference equal to the normal period, then look to see where the latency ends on that ring. Let the ring be colored in order of increasing latency red-orange-yellow-green-blue-violet-purple-red, just as it was in the exploration of circadian timing (Figure 3.9). As in that context, the coloring will help to show that the mere existence of even resetting has a surprising implication. This implication is not at all obvious, but neither can it be evaded. Strong resetting is incompatible with the common-sense notion that there can exist some measurable, reproducible latency following whatever combination of coupling interval and stimulus size we might choose. We can use color to illustrate this impossibility.

Painting Time in Rainbow Colors

The idea involved here is essentially geometrical, using the exact analogy between time in a cycle and hue on the color wheel. To cast the puzzle in a geometrical format, we need a geometrical way to represent the variety of possible stimuli, differing in magnitude from zero to some

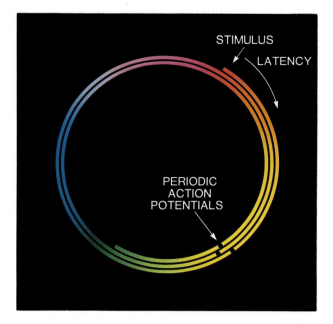

Figure 3.9: The time axis, measuring latency, starting the moment the stimulus ends, is rolled into a ring of circumference equal to one natural period in order to superimpose repeats of the action potential after the stimulus. Position around this cycle is color-coded, red coding the moment of the action potential.

physiologically realistic maximum of excitation (or of inhibition) and differing in timing throughout the cycle from one spontaneous beat to the anticipated next. The solution is simple: all those stimuli might be represented as points on a cylinder of graph paper that can be slit open and laid flat as a rectangle with coordinates *coupling interval* horizontally (the former circumference) and *stimulus size* vertically (Figure 3.10). This is the same rectangular layout used in the pinwheel experiments of Chapter 1.

Next, it will be useful to work into the same geometrical diagram some indication of the result of each such stimulus. We are focusing on just one aspect of the result: the latency (abstracting away the distinction between first, second, nth). Because it was decided above to represent this aspect of the latency by a color, the results are incorporated into the diagram of stimuli by coloring every point in the rectangle of possible stimuli (Figure 3.11) as in Figures 1.4 and 1.5.

How will the pattern of timing be colored? We can't really know in detail without measuring the result of every possible stimulus. But in the case of Figures 3.6 and 3.7, we do know the latencies following stimuli

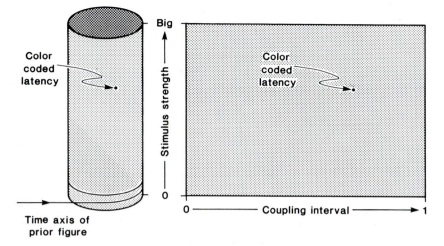

Figure 3.10: A series of experiments, each producing a coil of data like Figure 3.9, may be stacked into a cylinder in order of stimulus strength. The cylinder is here slit open to expose its surface as a rectangle like Figure 1.1, on which color-coded latencies are arranged in joint order of stimulus strength and coupling interval.

along the border of the rectangle. To anticipate: points along the rectangle's border wrap once around the color wheel. To see this, consider the four edges in sequence on Figure 3.11. The bottom of the rectangle, AB, wraps exactly 360 degrees: in odd resetting, latency decreases through one cycle as the coupling interval increases full cycle from left to right. The remainder of the rectangle's border continues, but without wrapping up further cycles or unwrapping the one already achieved, as can be seen by considering each side in turn. Along side BC latency varies in some way: we don't know exactly how it varies but it doesn't matter, as will be seen shortly. For the present it would suffice to suppose that the fixed coupling interval along BC lies within the refractory period of the pacemaker, so the stimulus has no effect. Then along edge CD we observe even resetting: latency increases and decreases (goes more purple then backs up through green again without scanning a cycle. Then along the descent DA it varies (but in reverse) just as it did along the ascent BC (at the same coupling interval but in the next cycle). So along the border ABCDA the colors traverse the complete color ring, as promised. We wind once around the color wheel in circumnavigating the rectangle once, so we say the "winding number" of color around ABCDA is 1.

To define a latency for every stimulus inside the rectangle is to paint every point with some color on the colored latency ring. Curiously, it is not possible to do so: inside the complete continuous ring of hues, there

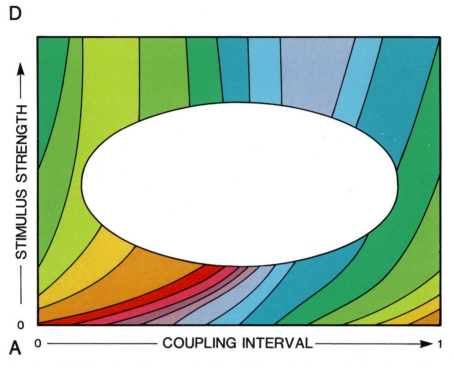

D   **C**

STIMULUS STRENGTH

0

A   0 ———————— COUPLING INTERVAL ————————▶ 1   **B**

Figure 3.11: The edges of the rectangle of Figure 3.10 are colored as required by sinus node data. Each point represents an inhibitory stimulus of some strength (vertically upward from 0) delivered at some time (horizontally from 0 to 1 period after action potential). The next beat's latency is indicated by coloring the point as revealed along the "strength = 0" axis AB, where coupling interval = distance from A and latency = distance from B. Red codes zero latency (immediate beat). Along CD (big stimuli at various times) the color only changes a little and changes back, following the even-resetting measurements of Jalife et al. 1983. As a result colors run once through the full cycle of saturated hues along ABCDA. Any attempt to color the whole rectangle (i.e. to determine latencies to diversely timed lesser stimuli) must reckon with at least a point of ambiguity somewhere inside.

must be a strictly discontinuous change of coloration or else some color not found anywhere on the border. Some point, then, must be hueless, e.g. grey. If you don't believe it, try to smoothly color the white space in Figure 3.11 with the hues already used in the figure.[2]

---

[2] Here we need to use words in a special way in order to proceed without unwanted verbosity. We call our cycle of distinct colors—the palette of colors used to encode time in a cycle—by the name "hues." A region is called "hueless" if it cannot be colored from that same palette without violating continuity. A "hueless" point need not be grey: it might be tinged with color. But it cannot be any color found on the encoding cycle. As a special case, the hues chosen may be the cycle of saturated colors from the rim of the artist's color wheel. The word "hue" can then be taken in the usual meaning: one of the three attributes of any color (hue, saturation, and brilliance). With this choice of meaning, "hueless" means "grey."

A Proof of Uncertainty

It is a curious fact (see Appendix) that you can proceed through the entire sequence of color wheel hues without backing up, and find yourself again at the beginning—rather like finding India by sailing stubbornly west. The peculiar consequence is that there must exist a color of uncertain hue—e.g. grey—and that some point inside the color wheel must be so colored. This may be obvious to painters but the rest of us have to resort to a topological theorem for certification that a hueless point is really necessary. The theorem is a generalized version of a familiar observation: a soap film on a wire ring cannot shrink out to its edges (on the wire ring) until some interior point is pricked; then the soap film can shrink away from that point, retracting to the perimeter. In its generalized version, this fact about continuous surfaces (that a central point must be omitted before the surface can retract to its boundary) is called the non-retraction theorem. To use it, we transform the phase-resetting problem until it resembles the soap film problem.

Outside the rectangle of Figure 3.11, draw a big ring and color it like a color wheel. Since color advances smoothly through a full cycle around both the rectangle's border and the exterior ring, the region between them can be colored with no difficulty, for example as indicated in Figure 3.12 (top): the left and right walls of the rectangle are identically colored, so they join point by point through isochrons of uniform color; the roof, consisting only of even-resetting segments along which hue changes uni-directionally then falls back again, connects to itself through concentric arcs of color; the floor, a smoothly colored full cycle, joins radially to same-colored points along the outer ring. This smooth coloring is displayed a little more symmetrically in Figure 3.12 (bottom) by distorting the rectangle to bring identical corners A and B together above the roof, stretching the full-cycle floor to resemble a full cycle.

But our objective was to color the remaining region continuously, inside the rectangle. (There is another constraint: the coloring outside the triangle of Figure 3.12 (bottom) must match that inside along the triangular interface; but we will see that it is hopeless anyway, even before trying to satisfy this additional constraint.) We can now use the non-retraction theorem to prove this impossible. The most direct approach is proof by contradiction. Suppose this disk could be colored, i.e. suppose every point could be assigned a color on the ring, with no point left ambiguous, with no discontinuities in the coloring as, indeed, the outer region can be. That coloring could then be used to assign a unique "home" on the ring to every interior point: all the red points could retract together to the one red spot on the rim, and so on. This contradicts the non-retraction theorem; thus there must be some interior point where distinct hues abut discontinuously and/or no hue can be assigned. By coloring the outer

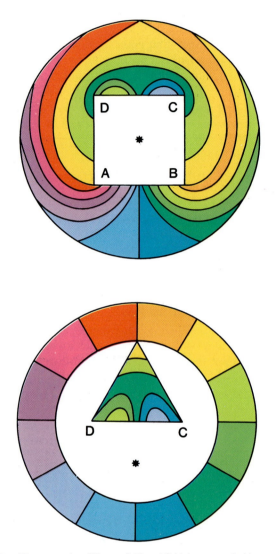

Figure 3.12. Top: The rectangle of Figures 3.10 and 3.11 is surrounded by a color wheel to show that the intervening space can be smoothly colored. The interior of the color wheel must contain a locus of discontinuity or a point of phase singularity, so it must be inside the rectangle. Bottom: The diagram is redrawn faithfully as far as topological essentials are concerned by bringing together points A and B (which represent the same experiment).

region as shown, we find that the hueless point is inside the rectangle.

That hueless point is a phase singularity: the winding number of latency around it is non-zero. How do we know? The winding number is 1 around rectangle ABCDA. Let a closed path shrink away from the boundary ABCDA. Its integer winding number remains inviolate as long as it moves through continuously varying color, because integers cannot change gradually (this fact may also be formalized as an index theorem [Glass 1977; Penrose 1979]). Ultimately this contracting noose encounters the discontinuity and fits around it, still with winding number $W = 1$ of latency. This paragraph alone constitutes sufficient proof that hue must somewhere become discontinuous or undefined, even without the argument of Figure 3.12.

The singular point might be bigger than mathematically required, but if it is a mere point, then it is a point where all hues merge—a hueless point. That stimulus inside the rectangle must have no hue. The latency of the beats following that stimulus must then be unpredictable or non-existent.

What is that critical moment between heartbeats, that singular coupling interval? If we could treat the whole heart as we treat its pacemaker node—as an oscillator, however complex its inner workings might be—then we might be tempted to ask, "Could the singularity somehow be related to the vulnerable phase?" Could this unpredictability of timing, this resetting to all phases simultaneously, provide the seed of the malignant asynchrony that we call fibrillation? Leaping ahead a few chapters, the answer appears to be yes, but in a way that would seem nonsensical or paradoxical if introduced here. A few more facts and ideas must first be considered.

The Essential Diagnostic

It is the purpose of theory to be as simple as possible. Especially in situations like this one, a shortcut is needed through fathomless complexities that might not seem relevant in the final analysis. But alluding to anything as arcane as a non-retraction theorem (illustrated by popping a soap film, verified with colored pencils!) might seem simple-minded to the point of absurdity. Let's stick with it, nevertheless, until we see what experiments it leads to. Meanwhile, it is important to be sure we know exactly what baggage we are carrying on this speculative trip.

What is the essential, indispensable physiological fact forcing us to the conclusion that heartbeat timing must be unpredictable after some stimulus that seems otherwise harmless? The exact way the color varies on the vertical sides of the rectangular box is unimportant for registering the total change of color in one circumnavigation of the box: color changes the same way but in opposite sequence as we ascend, then descend, the two identical sides. All that matters is how the color changes along the floor of the box (at zero stimulus size) and how it changes along the lid of the box. No matter what kinds of stimuli are used, no matter how a heart may react to them, the answer is always the same, and trivial, along the floor of the box: no stimulus, no resetting—the hue marches steadily though one full cycle around the color wheel. This means only behavior along the lid of the box is important. For present purposes, the only feature of "behavior" that concerns us is the way the hue moves around the color wheel as we traverse one full cycle of coupling interval, from one end to the other of the lid. Because the two ends represent the same stimulus, administered in either of two successive identical cycles, the color must be the same at both ends: it must accordingly traverse an exact integer number of full cycles along the lid, as it did along the floor. Thus all that matters now is that lid integer. If the integer is 1, as in the case of odd resetting, then the excess of one full cycle of hues achieved from left to right on the floor of the box is relieved from right to left along the lid and we have no problem. But otherwise, as in the present situation with even resetting (integer = 0), the excess is not relieved and there must be a hueless stimulus somewhere inside the box. The essential, indispensable physiological fact is the integer number of cycles of smooth phase change: if it is not 1, then latency is at least unpredictable and maybe even nonexistent following some smaller stimulus, rightly timed. Laboratory diagnosis that a phase singularity lurks somewhere thus boils down to making the distinction between one full cycle and no cycles at all. Such a difference is hard to miss empirically, and we therefore have in hand a very convenient and powerful diagnostic assay.

To summarize: a sufficient diagnostic criterion for this novelty is the mere observation of even resetting. Many experimental physiologists gathered such data, even decades ago, before comparable data were gathered (and also not recognized) from circadian clocks and biochemical oscillators. In the absence of some theoretical framework, the data were never plotted in such a way to make their topology obvious, and their counterintuitive implication was therefore consistently overlooked. But now it must be confronted: we must ask what singular arrhythmia really means in an oscillator composed of multitudes of (potentially) independently rhythmic fibers. We will do this in Chapter 5. But Chapter 4 comes first.

Can Neurobiologists Really Find Phase Singularities?

Each forward step in the construction of new science is initially only a dream without substance, a pseudopod of imagination protruding in search of attachment. Most are retracted and resorbed. Only when a solid fact is encountered can imagination harden and become a conduit for further dreaming. It is time now to seek that attachment, to make an empirical test of our inferences thus far to distinguish whether we are doing theoretical biology here or only theoretically doing biology. In *Tom Sawyer Abroad* Huck Finn remarks: "The trouble about arguments is they ain't nothin' but theories, after all, and theories don't prove nothin'. They only give you a place to rest on a spell when you are tuckered out buttin' around trying to find out somethin' there ain't no way to find out. And there's another thing about theories: there's always a hole in 'em sure, if you look close enuf."

Curiously, it is the very point of this argument that there must be a hole in any theory of even-type resetting that purports to reset the heartbeat in a completely continuous, orderly way without risk of arrhythmia. We seem to have here a model-free and theory-independent argument about the mutual timing of fibers that stimulate each other. Its conclusion is that synchrony must be impossible following some smaller stimulus than those known to evoke even resetting. That perfectly timed stimulus is a hole in the pattern of timing. Are such holes observed?

Let us plan some experiments. If this shortcut through the underbrush of cardiac physiology has anything to recommend it, it must invite laboratory demonstration of some consequences. Those consequences should be explicit, precise, and novel predictions. We begin by designing a precise experiment via quantitative calculations.

CHAPTER 4

Computer-Aided Design

of an Experiment

A good ... theory should not only correctly describe the current theoretical knowledge, but should also predict new results which can be tested by experiment. The further the predictions from the original experiments, the greater the credit to the theory if they are found to be correct. Thus observations of whether or not singularities actually occurred would provide a strong test. [Adams Prize Essay, Cambridge University, 1966]

The view has been expressed that singularities are so objectionable that if the ... equations were to predict their occurrence, this would be a compulsive reason for modifying them. However the real test of a ... theory is not whether its predicted results are aesthetically attractive, but whether they agree with observation. [Proc. Roy. Soc. Lond. A294, 511-521, 1966]

—Steven Hawking

The first effort to locate the predicted phase singularity in pacemaker cells was undertaken by Eric Best, then a student of aeronautical engineering, working on his Ph.D. [Best 1976] in the Department of Biological Sciences at Purdue University. Our idea was to defer confrontation with the imposing complexities of the human heart and first ask more modestly whether these topological notions apply even to the simplest periodically firing nerve cell. Even more modestly, it seemed prudent before tackling an actual nerve preparation in the laboratory to choose the presently best understood nerve preparation and ask whether these topological notions made sense in context of what electrophysiologists already understood about that nerve. But how?

Equations for a Squid's Escape

The best understanding of nerve dynamics is written in the language of differential equations. And the first well-understood nerve cell belongs to the squid. This long nerve, the squid's so-called giant axon, first lent itself to study by virtue of its architectural simplicity and its great size. It conveys the alarm message from the squid's brain to its jet-propulsion mechanism for escape. The way in which it does this was first formulated mathematically in 1952 by A. Hodgkin and A. Huxley, who won the Nobel Prize in Physiology in 1963 for their work. Their original equations (Figure 2.3) have been studied and refined for over thirty years by thousands of scientists.[1]

So well understood is this particular nerve that the Hodgkin-Huxley equations became the prototype for contemporary models of the heart's pacemaker tissues and their interactions. They are good candidates for a "thought experiment" in which the ideas about phase resetting developed in Chapter 3 can stand trial in the court of real electrophysiological dynamics.

The trial's outcome commanded suspenseful interest in 1975, because nothing like a phase singularity had ever been described in a neural pacemaker or in the supposedly equivalent equations. Could there still be something qualitatively new to discover in these equations after a quarter-century of feverish study? If not, then maybe there was something fundamentally wrong about using them (and their descendants) to describe pacemaker neurons, because something new seemed ripe for discovery in the laboratory. Many physiologists had measured even resetting since 1964, in many different kinds of pacemakers; this seemed inescapably to forebode observation of a phase singularity.

Then why had singularities never been noticed? Perhaps because they don't exist in reality. Or could it be just the needle-in-a-haystack dilemma? Maybe singularities really do exist but are so small they escape notice, as the planet Neptune did for many years and Nemesis perhaps does now. If so, then it should be possible to cajole the Hodgkin-Huxley equations into belatedly revealing their presence. The equations and the reality of singularities could then be tested by systematically seeking

---

[1] These equations and their descendants provide only a convenient quantitative description of the average time course of openings and closings among populations of ionic channels, proteins embedded in the cell membrane. For a thorough historical review see Scott 1975; for mechanistic interpretation in terms of molecular biophysics, see Hille 1984.

singularities right where they are predicted to be in living nerve preparations.

This is time-honored procedure. The planet Neptune was discovered by calculations. Unforeseen peculiarities in the orbit of Uranus, the outermost known planet at the time, seemed to need interpretation. Sticking exactly, literally, conservatively, to what they thought they knew of planetary dynamics, the theoretical astronomers John Couch Adams in England and Urbain Jean Le Verrier in France independently inferred that there must be a previously unsuspected object on the outer fringes of the known solar system, a point-like concentration of mass that might be big enough to see, even in the faint light of the Sun at that enormous distance. Coordinates were calculated in 1845 and given to careful observers at the Berlin Observatory. On the nights of 23–24 September 1846, conditions were good: they had a look, and there was the planet Neptune.

The neurobiologist's equivalent of unaccountable misbehavior in Uranus' path across the sky is even resetting in many kinds of pacemaker membrane. Can such an unmistakable diagnostic—the difference between a cycle and no cycle at all—be an observational error? We might begin to check by finding out whether even resetting is implicit in the governing equations of a particular pacemaker, the spontaneously firing squid axon. If not, either the equations are wrong after all, or squid membrane differs more fundamentally than has been suspected from the membranes of other rhythmically active nerves. But if even resetting is found in the equations, then a singularity must also be findable in the equations and then, knowing its coordinates, findable in the laboratory. Either way, the effort will be repaid.

Setting Up the Computer Experiment

The general who wins a battle makes many calculations in his temple before the battle is fought.
The general who loses a battle makes few calculations beforehand.
Thus do many calculations lead to victory and few calculations to defeat; how much more no calculations at all!
It is by attention to this point that I can foresee who is likely to win or lose.
—Sun Tzu, *The Art of War* (500 B.C.)

In such an enterprise, one rarely has to start from the very beginning. Usually something similar has been tried before, somewhere, if one can only find it, and that earlier effort need only be adapted to the new purpose. In this case all the pieces were conspicuously at hand. Even resetting and singularities had been sought and found in half a dozen circadian clocks [Chapter 1 and Winfree 1980], in oscillating sugar metabolism [Winfree 1972c, 1980], and in the dynamics of water transport in seedlings [Johnsson 1976]. The topological theory motivating those measurements had been formulated and widely published, including the pinwheel experiment and the singularity trap protocols [Winfree 1980]. They had been used successfully and could be quickly adapted to Best's new context. Existing data on pacemaker response to a single stimulus had been reinterpreted to reveal even resetting. The equations of spontaneously rhythmic nerve membrane were known to be of the same type as those implicated in the circadian clocks, etc. Reliable algorithms had already been written and refined for solving arbitrary sets of dynamical equations. The pieces had only to be put together.

Putting ourselves in the position of Best in 1975, then, the thing to do is to choose a good solver routine from those available, enter the membrane equations, and test the program in simplified cases, to verify that the outcome is exactly what it ought to be if everything works right.

With some confidence that the computer is actually doing what we want, we can enter instructions to simulate the intended (pinwheel) experiment, whose outcome is *un*known. In particular, the axon is first to be space-clamped, which means that its full length is compelled to keep the same electrical potential. In a living nerve, this is arranged by impaling the nerve on a long wire that runs its full length. The wire, being a nearly perfect electrical conductor, is always at uniform potential, and this fact dominates the electrical potential inside the axon. In a computer simulation, the space-clamp is arranged by simply declaring that there are no potential differences or currents along the axon's length: only between the (electrically uniform) inside and (electrically uniform) outside will there be potential differences and the corresponding current flow.

Next, the space-clamped axon must be made to fire rhythmically in the computer. This can be arranged by continually providing a tiny depolarizing current across the membrane.

Then the simulator must be programmed to deliver a simulated electrical stimulus at some coupling interval after a firing. The stimulus is an instantaneous erasure of one number and its replacement by another: the membrane potential is instantaneously offset by some physiologically realistic number of millivolts, positively or negatively. In the living nerve this can't be done quite instantaneously. It is done in the laboratory by driving a big current through the membrane for a very short time until the required potential difference accumulates. In the living animal, it is done by chemical changes in the membrane (induced by neurotransmitter release from an adjoining nerve) that open protein pores to a flow of ionic currents. Thus we have three different meanings of "stimulus": a fixed offset of voltage, regardless of present voltage and membrane resistance (used by Best); or a transfer of a fixed number of charges, regardless of present voltage and resistance, thus imposing different voltage stresses on the membrane depending on its conductivity at the moment (used in all other calculations and experiments discussed in this chapter); or a brief change of membrane resistance whose effect in terms of current and voltage doubtless depends on present voltage and resistance (used by cells). For our purpose of establishing the existence of even resetting and a phase singularity, the difference should not matter; therefore we choose the simplest: the voltage offset.

Having set up the stimulus, we program the computer to record its aftermath, resuming normal computations after the instantaneous offset of electrical potential. We instruct the computer to print out the latencies of subsequent firings alongside the chosen stimulus strength, in a column above the chosen coupling interval. It turns out that firings still follow one another at almost exactly the interval that was normal before the stimulus, so we need tabulate only the first of them: the first post-stimulus latency.

Finally, the simulator is programmed to scan the pacemaker cycle with such tests, repeating the tests at many stimulus sizes, both excitatory and inhibitory.

Making the Results Presentable

When the computed latencies are tabulated by coupling interval and stimulus size, they make a big rectangular array. It looks as indigestible as any big corporation's annual financial report, but is much more transparently structured. You most likely learned one such tabulation by rote in school: the multiplication table. If you were resourceful about finding memory aids, you probably examined the table for hidden structure; you noticed, for example, that the numbers get bigger as you go down or to the right. You could have painted hyperbolic curves on your multiplication table to link all the products that come to about 20, adjacent to a parallel corridor of products that come to about 30, and so on.

The same procedure helps to make sense of this table of latencies: we sketch contour lines (isochrons) through similar latencies. Each contour line depicts the combination of depolarizing stimulus size (on a vertical scale from 0 to 53 millivolts, as an instantaneous offset) and coupling interval (on a horizontal scale of one full period) that resets the squid giant axon pacemaker to the same latency (firing time after the stimulus), according to calculations from the Hodgkin-Huxley equations. Then we label the contours: one represents all those combinations of coupling interval and stimulus size that result in firing after 5 milliseconds; an adjacent, almost parallel contour connects all the 6-millisecond latencies, and so on. What matters here is not the exact number of milliseconds, but the latency in comparison with the normal duration of a cycle; the contours are simply labeled from 0 to 20 by twentieths of a cycle (Figure 4.1). Firing (here signaled by peak voltage rather than by initial depolarization) is represented by "20" = "0"; firing continues at unit intervals after the stimulus. Notice this contour bounding the rectangle and appearing again at the top at coupling interval about 1/7, then swooping down to the right. This is the nominal boundary of the "absolute" refractory period: prompt firing (short latency) can only be obtained by stimuli that are later or more depolarizing than this locus. (It was stated in error in Winfree 1981 that an artifact of these computations masked the expected refractory period. But there is no artifact; this contour delineates the expected region. Note also that "refractory," as used in neurobiology, does not mean "unresponsive," but only that the response is not a premature action potential. Thus any phase at which greater stimulation fails to advance the process or puts the oscillator on an earlier [delayed] isochron is "refractory.")[2]

Actually, the written labels are superfluous, because each latency contour identifies itself where it starts on the zero-stimulus axis: the contour crossing the axis one-third of a cycle from the right end must be the contour of latencies = 1/3 cycle, because there is that long to go until next firing after a negligible stimulus administered 1/3 cycle before full cycle. By the same reasoning, this is the contour of new phase = 2/3 cycle. Each contour labels itself by touching the zero-stimulus axis that many twentieths before the next firing in the unperturbed control.

Not only are the labels on the contours superfluous, but it is somewhat misleading to have discrete contours at all, as though there were none

---

[2] The notions of "advance" and "delay" (Figure 3.3) turn out to be ambiguous unless one of several arbitrary definitions is specified. If behavior after the stimulus is not a rigidly shifted replica of normal behavior but is distorted, then the adjectives cannot be applied in any exact sense without first deciding which event to time: maximum depolarization regardless of the voltage? the inflection point of the voltage trace or the moment of most rapid depolarization? These arbitrary choices of convention can make a qualitative difference in the outcome. But if there is no distortion, Figures 4.1 to 4.5 and 4.9 still show that other ambiguities can arise. A stimulus that, when prolonged or applied more vigorously, results in greater latency, may be called "delaying"; but this differential test can have different results depending on the duration or strength. If the differential test is not done, "advance" (rather than "advancing") or delay (rather than "delaying") may be reckoned from the absolute latency, possibly with a different conclusion. My choice is to avoid those adjectives wherever possible. They are not in the Glossary.

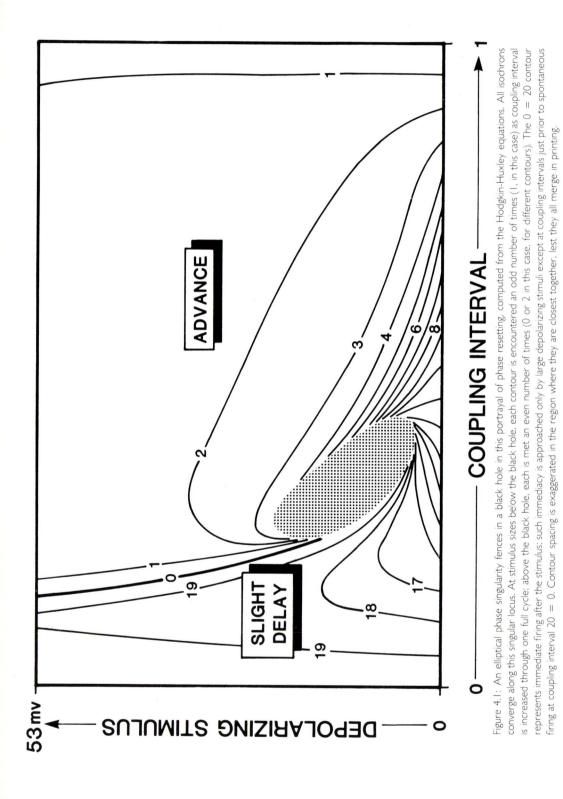

Figure 4.1: An elliptical phase singularity fences in a black hole in this portrayal of phase resetting, computed from the Hodgkin-Huxley equations. All isochrons converge along this singular locus. At stimulus sizes below the black hole, each contour is encountered an odd number of times (1, in this case) as coupling interval is increased through one full cycle: above the black hole, each is met an even number of times (0 or 2 in this case, for different contours). The 0 = 20 contour represents immediate firing after the stimulus; such immediacy is approached only by large depolarizing stimuli except at coupling intervals just prior to spontaneous firing at coupling interval 20 = 0. Contour spacing is exaggerated in the region where they are closest together, lest they all merge in printing.

COUPLING INTERVAL → 1

0 →

53mv ← **DEPOLARIZING STIMULUS** → 0

Figure 4.2: The isochrons of Figure 4.1 are colored to become "isochromes" according to the color code (red = 0) used throughout this book. (The corresponding figure of Winfree 1983 was incorrectly printed in *Scientific American*: left edge should have been red.)

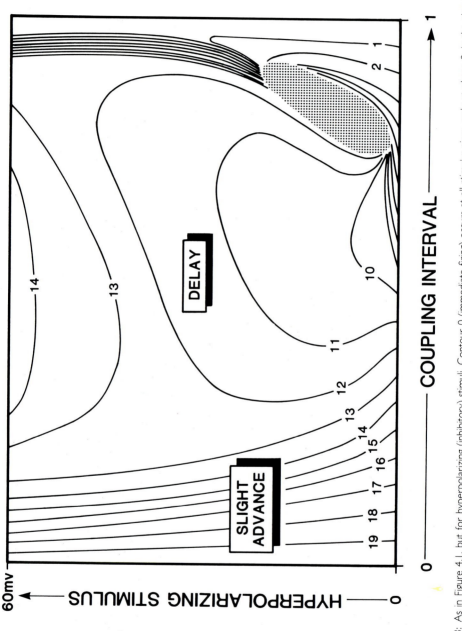

Figure 4.3: As in Figure 4.1, but for hyperpolarizing (inhibitory) stimuli. Contour 0 (immediate firing) occurs at all stimulus sizes, only at phase 0, in the absolute refractory period chosen to bound the box at left and right.

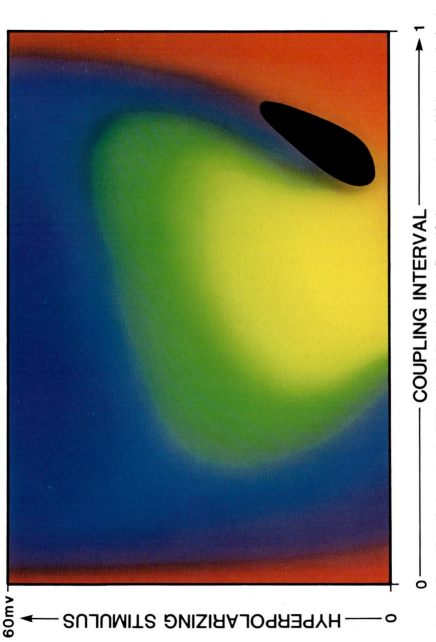

COUPLING INTERVAL

0 — 1

HYPERPOLARIZING STIMULUS

0 — 60mv

Figure 4.4: As in Figure 4.2, but for hyperpolarizing (inhibitory) stimuli. Note that just as in Figure 4.2, there are regions in which increasing stimulus size moves the pacemaker to later colors (delaying, as might be expected here, but a little counterintuitive for the depolarizing stimuli of Figure 4.2), and other regions in which the movement is to earlier colors (advancing, perhaps paradoxically here, but expected in Figure 4.2). (The corresponding figure of Winfree 1983 was incorrectly printed in *Scientific American*: left edge should have been red. Here, in the region of steep color change issuing upward from the singularity, the computed gradient is much steeper than shown: if it were done to scale, the compressed half-cycle of color bands would be too narrow to print.)

between them, when actually all intermediate values of latency are equally important and should be equally represented. A procedure we have used before recommends itself: to color-label the contour lines, then let the colors spread and blend, dissolving the discrete lines (Figure 4.2). If you prefer contour lines, you can still trace them anywhere along the directions of constant hue as in Figure 4.1; many people do. To borrow a pun from Stuart Kauffman of the University of Pennsylvania Medical School, the colored isochrons are called iso*chromes*. Cardiologists compromise, using the terms isochrones or isochronal lines.

Either way, have a look at the latencies computed after a substantial depolarizing (excitatory) stimulus (Figures 4.1, 4.2): as we vary the coupling interval from left to right through a full cycle, the latency scarcely varies. Firing is immediate (latency negligible, new phase always reset near contour zero, color red). The result of a gentler stimulus depends more on when it is given. In any case, every color is met with twice in our scan from left to right: the Hodgkin-Huxley equations do, as foreseen, describe even resetting.

The foregoing pictures reflect the action of excitatory stimuli such as electrical currents. Figures 4.3 and 4.4 show the same computations but with inhibitory (hyperpolarizing) stimuli like those that, in the case of the heart, arrive along cholinergic branches of the vagus nerve. Figure 4.5 plots the hyperpolarizing (upward) and depolarizing (downward) parts together on one vertical stimulus axis as though membrane polarization were perturbed through an intracellular microelectrode.

Black Holes in the Color Field

The vital feature revealed in this latency chart is that all the colors in either half do indeed converge toward a phase singularity, as anticipated. But why so big? Figures 3.11 and 3.12 required only a point of huelessness, but here we find an elliptical ring of phase singularity and inside it, blackness.

This hollowness is no accident. It was deliberately arranged by delicate adjustment of electrical conditions to make the membrane's resting potential stable while it also has a stable cycle. In this condition the resting membrane does not spontaneously begin rhythmic firing; it remains inactive until temporarily nudged far enough off equilibrium. Only then will it continue to fire rhythmically. Some such phenomenon was explored in the early 1970s [Ferrier et al. 1973; Wit and Cranefield 1976]; cardiac physiologists call it triggered activity. From its rhythmic state the membrane can be made quiescent again by somehow bumping the membrane to any of a range of states from which it will relax back to the original steady state. The edge of that basin of attraction is the phase singularity;

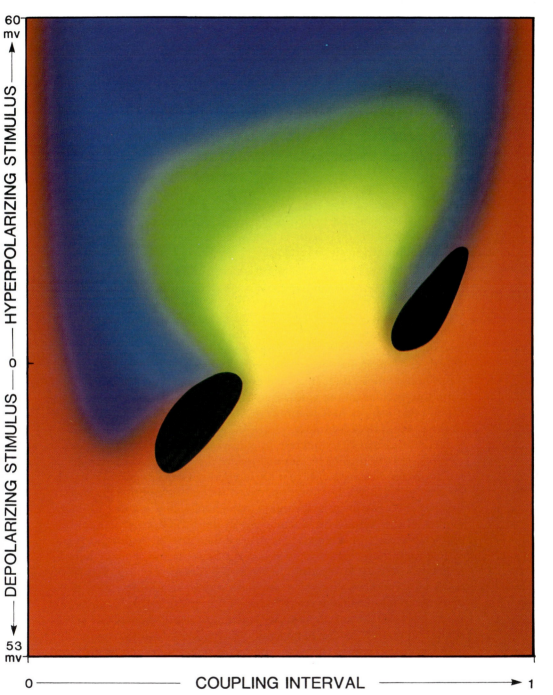

Figure 4.5: Figures 4.2 and 4.4 are combined in one diagram with hyperpolarizing stimuli plotted upward and depolarizing stimuli plotted downward. Notice that every isochrome (locus of same new phase and same latency) arcs from one singularity to the other. This figure and Figures 4.1–4.4 are adapted from Best 1976 and/or Best 1979. (The corresponding figure of Winfree 1983 was incorrectly printed in *Scientific American*: left edge should have been red. Here, in the region of steep color change issuing from each singularity, the computed gradient is much steeper than shown.)

if the basin is big, not a mere point, then the singularity, its border, is also not a mere point but a ring.[3]

In our first search for the predicted singularity, Best and I felt it might be overlooked unless it were bloated to conspicuous size by a tumorous black hole. A black hole is hard to overlook because following any stimulus inside it, rhythmic calculation stops. In 1975 Best thus sought to make the theoretically anticipated singularity as conspicuous as possible so that if it were not found in this computer experiment, it could not be argued that it was missed because it was so small. But this required delicate adjustment of a constant leakage current across the membrane to make the membrane's capacity for self-sustained rhythmicity coexist with capacity for stable quiescence. In more typical situations, quiescence is *un*stable and the singularity is a mere point, a needle of seemingly little practical concern in the haystack.

This brings us to a subtlety about phase singularities that seldom fails to confuse: they exist and organize the system's timing regardless of their size. Size is not the most important thing about a phase singularity. In many models it is a mere point in the rectangle of stimuli and would never be observed exactly, even by the most extraordinary luck. The important thing is the arrangement of timing that only occurs around a singularity. The remainder of this book will present abundant examples. The "black holes" that arise in the next and following chapters derive from this circular timing gradient in a *spatial* context; in no way do they depend on the black hole that may or may not lurk within the singularity in the space-clamped context.

Nonetheless, it is of interest to know what happens in the space-clamped situation after a stimulus inside the black hole. The colorful latency diagram shows only that the latency will be ambiguous along the borders of the black hole and undefined within. It might be ambiguous because rhythmic firing will resume immediately, but (because of uncertainties in the stimulus or fluctuations in the oscillator's dynamics) unpredictably reset in timing. Or it may resume after a random interlude of low-amplitude fluctuations. Inside the black region, latency may be undefined because firing continues with a different period or waveform or because there won't even be any more rhythmical firings. That is the outcome in this particular contrived case: the Hodgkin-Huxley equations desist from their prior rhythmicity, relaxing toward their stable steady state (Figure 4.6). But more generally, the way in which the arrhythmia manifests itself depends delicately on circumstances and can be arranged

---

[3] The phase singularity is only the point or locus where phase becomes discontinuous because all isochrons come arbitrarily close together; inside that locus (if it is not just a point) phase is simply undefined. The "phaseless set" consists of the phase singularities together with regions of undefined phase.

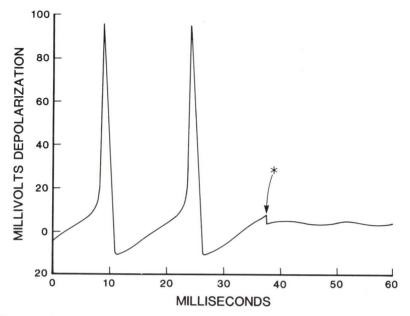

Figure 4.6: Computed 15-msec pacemaker rhythmicity in the squid giant axon is terminated by a minuscule hyperpolarizing stimulus just before the spontaneous firing. This stimulus lies inside the black hole of Figures 4.3 and 4.4. From Best 1976.

in any of the above ways and perhaps others. As we will see, living tissues are ingeniously diverse in their arrhythmias, particularly if the oscillator has spatial extension. But what is robust and inescapable in this story is that some kind of arrhythmia will occur where the isochromes are forced to a hueless convergence, and some may also occur inside that locus if it is more than a mere point.

The isochromes are forced to converge (and so to violate our presumption of continuity) by the topological arrangement of colors around the perimeter of either box (Figures 4.1 and 4.2 for the box of depolarizing stimuli, and Figures 4.3 and 4.4 for the box of hyperpolarizing stimuli), as discussed near Figures 3.11 and 3.12. We see in these computations just what was foreseen in that discussion of even and odd resetting. Consider a circumnavigation of the box. Along the floor of the box (no stimulus), latency smoothly and trivially decreases through one cycle as the coupling interval increases full cycle. Wherever we choose to erect the side walls of the box (one cycle apart), the color change (latency change) accumulated as we ascend one wall will be exactly canceled in descending the other; in fact we might as well let the walls curve along isochromes at each end, so that the change is zero in both cases. Thus only the lid of the box contains information specific to the experimental system or equations in hand. In this computer experiment our heritage of understanding about squid membrane biophysics predicts that along the lid of the box latency will merely increase and decrease back again without progressing (net) through any full cycles of color change: this is even resetting. The consequence, as before, is the entrapment of a full cycle of isochromes inside the box. They must come together at some stimulus (possibly part of a closed locus of stimuli) inside the box; that stimulus cannot have any hue or it would be on some identifiable isochrome. And this occurs, as in Figure 4.5. The colored isochrons connect one singularity to the other in orderly fashion, wrapping a rainbow of color around each hole. The upper black hole tells us that a rightly timed inhibitory stimulus will inhibit forever—if it is not too big. The lower black hole tells us, perhaps paradoxically, that even an excitatory stimulus will do the same. Neither phenomenon, in 1975, had ever been observed. Is it really true that within each black hole lurk utterly hueless stimuli following which it is apparently impossible to predict the latency of spontaneous firing? Anyone who proclaims an "impossibility," whether in politics or science or technology, usually becomes a laughingstock within appallingly few years. In this case, however, it seems truly impossible. It is time for another experiment.

Living Squid Membrane

If topology can tell us there must be at least a point of arrhythmia, maybe topology can also guide us to it. Hole in hand, we might then begin to understand something of the physiological meaning of its singular border. But theoretical biology is a precarious high-wire act, disposing its practitioners to lightheadedness, and they tend to spend most of their time falling off. As Michael Faraday once observed in a similar situation, "all this is a dream. Still, examine it by a few experiments. Nothing is too wonderful to be true if it be consistent with the laws of nature. And in things such as these, experiment is the best test of such consistency" [letter, 19 March 1849].

At this stage we are finished with reasoning and simulations; it is time to wonder whether their consistently peculiar outcome is to be taken seriously. The first one to ask is the squid itself. Because the Hodgkin-Huxley equations were tailored especially for squid nerve, it is possible to take the computer experiment quite literally as a quantitative design for an actual experiment. John Rinzel at the National Institutes of Health undertook the experiment with Rita Guttman and Stephen Lewis at the Marine Biological Laboratory in Woods Hole, Cape Cod, in the summer of 1978 [Guttman et al. 1980].

At Woods Hole fresh squid are available daily from boats that dock

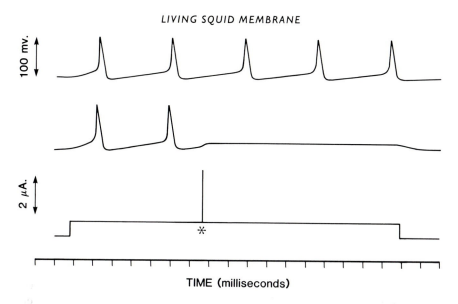

Figure 4.7: As described in the text, 4-msec pacemaker rhythmicity in the squid giant axon is terminated by a mid-cycle depolarizing impulse. The upper trace is a control given only the bias current, not the spike at time *. This stimulus may represent the black hole of Figures 4.1 and 4.2. From the experiments of Guttman et al. 1980.

just across the street. Sunlight and salt air pervade the labs; scientists plan experiments and ponder academic politics at the café tables, elbow to elbow with curious tourists. The squid experiment was implemented straightforwardly. Space-clamping an axon from a fresh live squid, Guttman made it fire rhythmically in a low-calcium medium by maintaining a constant slight current through the membrane. Electronic gear detects both the stimulus and the membrane's firing, displaying them both on a video screen as shown in Figure 4.7. In the bottom trace the biasing stimulus has been turned on, maintained for 17 milliseconds, then turned off. The top traces show that as long as the bias remains on, the nerve can fire periodically. This continues in the top trace, a control experiment subjected to no further stimulus. But at time * in the bottom two traces a stimulus in the *black hole* range was given: the right amount of excitation at just the right time in the cycle. Afterward the middle trace differs from the upper, unperturbed control trace only in that the nerve never fired again: its latency is undefined.

There is nothing wrong with the nerve; the annihilating stimulus has not damaged it. It is dislodged from this stable equilibrium to resume its rhythmical firing when the biasing current is turned off and then back on again. But until some large stimulus intervenes, the membrane's latency remains indeterminate, as predicted by the disappearance of all isochromes along the multihued fringe of this hueless region.[4] Actually, the isochromes converged to two hueless areas: one for excitatory stimuli, accessible by depolarizing the membrane just after it has repolarized, and another for inhibitory stimuli, accessible by hyperpolarizing the membrane just before its spontaneous rapid depolarization. Both were found in the experiment with living nerve.

What is the significance of this? Our purpose was to inquire whether biophysicists understand pacemaker mechanisms well enough today to give a reasonably complete account of phase resetting. This is still unknown: no phase resetting was measured on the squid pacemaker. But our topological tools and the quantitative equations of electrophysiology at least vindicated themselves as far as successfully predicting a new and surprising phenomenon. This example was not further pursued, but it gave courage to several groups to pursue related cases by the same methods.

For example, Eric Peterson (whose shattering of the mosquito's sleep/wake cycle was mentioned in Chapter 1) and Ronald Calabrese in the Harvard Biological Laboratories also probed the tiny heart of a leech with depolarizing stimuli [Peterson and Calabrese 1982]. Both odd and even resetting styles appeared clearly in response to small and large stimuli, respectively. And a stimulus of intermediate strength (only), if and only if applied at a critical moment, disorganized the heartbeat. In this case the phase singularity was a mere point without interior. A little while after the unique annihilating stimulus, normal beating always recovered from this confusion. But in contrast to the distinct reproducibility of timing after any other stimulus, this spontaneously recovered rhythm was unpredictably phase-reset.

Is a Heart Fiber Anything Like a Squid Nerve?

"Data, data, data," he muttered; "I cannot make bricks without clay."

—Sherlock Holmes, in "The Adventure
of the Copper Beeches"

While neurobiologists were preparing the squid experiment at Woods Hole, cardiac physiologists prepared a second test at the Masonic Medical

---

[4] Strictly speaking, the topological arguments require only that there be at least a point at which all isochromes converge to ambiguous hue. The locus of convergence in this particular case is more extensive than a mere point, but it is not the entire black area; it is only the hole's border. By painting the insides hueless, the topologist only takes a guess that beyond the timeless horizon (the multihued border) there are no colors at all inside (i.e. the oscillator will not recover to the usual cycle).

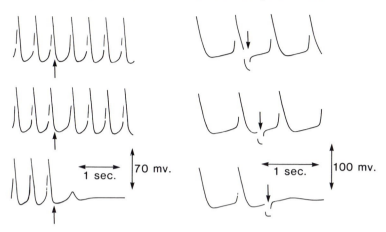

Figure 4.8. Left: Analogous to Figure 4.7, a suitably timed stimulus (a brief subthreshold depolarizing current, 50 msec at 4.2 μA at phase about 0.36) annihilates the 400-msec pacemaker rhythm in the sinoatrial node of the cat. Right: In a second experiment a suitably timed stimulus (a brief hyperpolarizing current, 100 msec at 6.4 μA at phase about 0.67) annihilates a 700-msec pacemaker rhythm, also in a kitten's sinoatrial node. Adapted from Jalife and Antzelevitch 1979 with permission from the authors and *Science*.

Research Laboratory in Utica, New York. Jalife's measurements of resetting in cardiac pacemaker cells had revealed both even and odd styles. This implied something that had never before been observed: that a single stimulus might unpredictably reset the timing of a vertebrate's sinoatrial pacemaker, if and only if it were just the right size and arrived at just the right time. Maybe it would even turn off the pacemaker. If so, could this be related to clinically important arrhythmias? What size stimulus would be required, and how exactly must it be timed? Was this a practically important possibility?

Why not seek the singularity in a cardiac pacemaker, e.g. the kitten's sinoatrial node? It would require delicate adjustment of the timing of a physiologically reasonable stimulus, much as in the squid experiment. Jalife and his collaborator Charles Antzelevitch, now director of the Masonic Medical Research Laboratory, tried almost a decade ago [Jalife and Antzelevitch 1979], using neurally released acetylcholine to briefly open the node's potassium channels as in normal regulation of the pacemaker rate. They discovered a subthreshold (i.e. delicate, physiologically commonplace) stimulus that caused abrupt arrest of a lifetime of previously uninterrupted pulsation (Figure 4.8), but only if exquisitely timed. In fact, they discovered both of the predicted singularities: an excitatory one about 0.36 period after action potential upstroke and, at about 0.67, an inhibitory one. The annihilation was sometimes permanent, suggesting a black hole inside the singularity; and sometimes it was ephemeral,

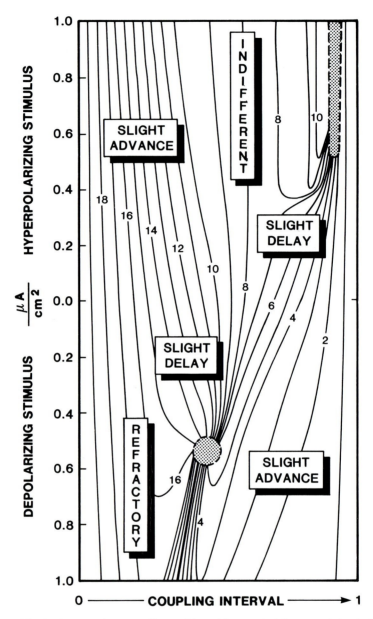

Figure 4.9a: Isochrons analogous to Figures 4.1 to 4.5 computed from an electrophysiological model of the sinoatrial node, assuming a 0.7 μA/cm² current bias, using stimuli of 50-msec duration at current densities indicated vertically. Coupling interval increases from 0 to 1 horizontally; 0 (left and right edges) is the midpoint of action-potential upstroke. The computed phase changes more abruptly near the regions of convergence than appears from this smooth contouring: the congested region at the upper right had contours too close together to print; just how close is unknown, as the computations were not pursued in enough detail to resolve ostensible discontinuity in the computed first latencies. (First latencies, in contrast to asymptotic new phase, typically show discontinuities [Kawato and Suzuki 1978; Kawato 1981].) Better resolved, it might resemble the singularity-cum-bundle of isochrons in the lower middle. Adapted from Reiner and Antzelevitch 1985 with permission of the authors and the *American Journal of Physiology*.

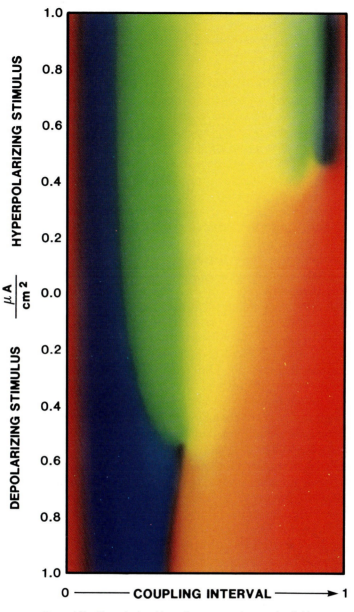

Figure 4.9b: The calculated latencies portrayed as a color field.

eventuating merely in unpredictably reset phase. In either case, normal rhythmicity could be restarted by a brief pulse later: the membrane had not been damaged.

What about the whirlpool of color surrounding the black hole? Before undertaking a search for the singularity, Jalife and Antzelevitch had spot-checked the timing of heartbeats that were merely phase-shifted by ordinary stimuli. Odd resetting appeared in response to smaller stimuli, and even resetting in response to stronger stimuli. As in Figures 1.2 to 1.5, this pattern when presented as a three-dimensional time crystal amounts to a screw surface surrounding the singularity. Following the analogous story in Chapter 1, Jalife had plotted it this way on a stack of lucite sheets as a guide in seeking the singularity of the kitten's sinoatrial node. It proved a reliable guide.

Much is now understood of the ionic mechanisms underlying pacemaker rhythmicity in the sinoatrial node. They differ markedly from the corresponding mechanism in squid axon. For example (worthy of particular attention in Chapter 5) the node is a spatially extended population of coupled cells, each independently periodic in its activity, but normally synchronous with its neighbors. Even more strikingly, in healthy nodal cells the fast sodium channels are inactive. (These are the membrane proteins that mediate rapid sodium ion influx during the regenerative depolarization that abruptly initiates action potentials in the squid giant axon and other nerves.) Instead, the upstroke in nodal cells is mediated by a much slower influx of calcium and sodium. Nodal cells have at least five distinguishable ionic channels, all with slow kinetics, and thus their action potentials are more rounded and propagate relatively slowly.

Quantitative models of sinoatrial membrane kinetics can be assembled in the style pioneered by Hodgkin and Huxley [Yanagihara et al. 1980]. It appears that ion pumps and concentration changes cannot be ignored in this system [Brown et al. 1984ab; Noble and Noble 1984]. There remains much disagreement about even the best models [Noble 1984], few of which—for the sinus node or any other cardiac oscillator—are reliable in predicting phase shifts [Reiner and Antzelevitch 1985; Guevara 1984, thesis 3-33; Michaels et al. 1984; Chay and Lee 1984]. Nonetheless, one candidate model has recently been used to compute sinus node phase-resetting behavior in response to positive or negative current pulses of various sizes applied at various times [Reiner and Antzelevitch 1985]. Odd and even resetting emerge, as can be seen in the contour map of Figure 4.9a. The format is the same as in Figure 4.5. The isochrons (connecting stimuli that give the same latency or new phase) in Figure 4.9b are colored in to become "isochromes." Their staircase succession along the zero-stimulus axis shows the expected odd resetting in response to slight stimuli. At large depolarizing stimulus levels, in contrast, one encounters only some contours, each twice, while progressing horizontally at fixed stimulus intensity through a full cycle of coupling interval. This is even resetting as in Figure 4.5. Thus an excess of one full set of contour lines enters the lower box of depolarizing stimuli (below the stimulus = 0.0 horizontal) and does not come back out. They converge at a phase singularity. Notice that the calculated depolarizing singularity is found somewhat after repolarization (phase, or coupling interval, 0.47), as in the experiment of Figure 4.8 (0.36). If the elongated obscure region in the upper right corner of Figure 4.9 hides a hyperpolarizing phase singularity just before firing (as Reiner and Antzelevitch assert more positively than I show it here), then calculation (phase 0.9) is again coming out roughly[5] as observed (phase 0.7) [Jalife and Antzelevitch 1979]. Though latencies are poorly resolved near the two zones of convergence (stippled or grey areas in Figures 4.9a or 4.9b), the phase singularity hidden within may be as small as topological necessity allows: a mere point rather than the border of a black hole. In fact Reiner and Antzelevitch were not even able to find it in their basic model; Figure 4.9 was calculated from a membrane chronically hyperpolarized by a current about as large as the singular stimulus. As in many other models and experiments with diverse biological clocks, "annihilation" is therefore unstable to arbitrarily minute fluctuations and cannot be achieved exactly in any real experiment, even a numerical experiment, with stimuli of inexact timing and magnitude.

The important feature for our purposes is that around the singularity, new phase (or latency) values run through a complete cycle. The contour map in that region thus represents a spiral staircase, the core of a screw surface with a helical boundary in each unit cell of the time crystal, much as Jalife plotted on lucite in anticipation from limited experimental data.

Turning to a system that lends itself to exhaustive and reproducible measurements, we can actually see this screw surface. Wilbert van Meerwijk of the University of Leiden Physiology Department made Figure 4.10 [van Meerwijk et al. 1984]: a single unit cell of the now-familiar

---

[5] Figure 3.7 shows the measurements of Jalife et al. 1983 on sinoatrial node perturbed by a hyperpolarizing stimulus. Their data fall out in even format, but not exactly as in Figure 4.9. The Reiner and Antzelevitch 1985 calculation shows an insensitive period (contours practically vertical, latency the same at large M as at small) right after the action potential (at early coupling intervals). In terms of the contour map, this refractory region is bounded by a vertical segment of the 0 contour at coupling interval 0, and another segment that dives into the singularity from extreme depolarizations. In terms of the resetting curve, refractoriness corresponds to a segment of curve along the control diagonal passing through the corner of each unit cell. In Figure 3.7 this segment starts too early and rides at higher latency. The reason may be that the stimulus used was not directly electrical, but an electrical activation of acetylcholine release to receptor channels, whereas Reiner and Antzelevitch simulated a discrete current pulse of 50-msec duration.

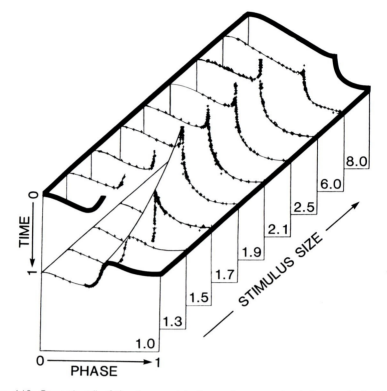

Figure 4.10: One unit cell of the time crystal of a cardiac pacemaker ball, perturbed with a depolarizing current pulse of 50-msec duration, at current as indicated on each resetting curve. "Phase" means the same as "coupling interval." "Time" is measured from the pre-stimulus action potential, so it represents latency *plus* coupling interval, in our notation. This skews the plotted surface 45° relative to our convention, but its topological features—the screw shape and the helical boundary—are independent of such details of presentation, and are evident here. As in Figures 4.1 to 4.5 and 4.9, the region of steep phase change near the singularities is quite narrow in the old phase direction but quite elongated in the direction of stimulus magnitude. From the experiments of van Meerwijk et al. 1984.

time crystal, but this time obtained in a curious way that permits greater precision and reproducibility.

To digress for a moment, let me describe the "curious way." It was first attempted successfully by Steven Scott in his Ph.D. thesis at S.U.N.Y Buffalo [Scott 1979]. Dropping the heart of a chicken embryo into a warm bath of dilute digestive enzyme (trypsin), Scott dissolved the collagenous glue that holds the heart's cells together. Transferred to a more hospitable bath, they reaggregate into minuscule balls of cells, each beating like a microscopic heart, much more regularly than any isolated cell. The regular rhythm is easily disturbed by an electronic pulse delivered through a submicroscopic glass syringe. With a computer sizing and timing the pulses and recording the shifted rhythm of beating, data accumulate swiftly. (In 1966 I attempted this experiment in the laboratory of Robert DeHaan; but impaling single cells unstably, and without computer assistance, data definitely did not accumulate swiftly.) Using fixed-current stimuli substantially less than 100 nA-msec on pacemaker balls around 100–150 microns in diameter, Scott eventually obtained astonishingly reproducible and smooth resetting curves of odd type. In the higher stimulus ranges, they developed an apparent discontinuity. It was not feasible to inject sufficient current to discover the even-resetting range.

Van Meerwijk et al. [1984] contrived a similar arrangement, but were able to inject several hundred nA-msec of depolarizing charge (at current densities averaging less than 1% of the normal peak inward sodium current). They found both resetting types (odd type below 1.6 nanoampere for a 50-msec pulse, and even type above that) fitting together three-dimensionally in a screw surface. Just one turn is shown in Figure 4.10; the later beats in each experiment delineate replicas above to make a full screw; and experiments using stimuli in later cycles would duplicate the screw again and again to the right: the edges of this composite surface fit smoothly together where the unit cells meet. What happened at the singularity? The microscopic hearts were unable to lie still for more than a few normal intervals: they always resumed beating but (after the singular stimulus) were unpredictably out of step with their own former rhythms.

Similar results were obtained at the same time with a similar preparation, using somewhat different experimental methods, at McGill University Medical School [Guevara et al. 1986]. Guevara also observed both odd and even resetting at depolarizing stimulus intensities comparable to van Meerwijk's. But as noted in Box 3.A, they were separated by a range of stimulus intensities (roughly twofold) within which the resetting curve is discontinuous, much as observed by Scott. Like Scott's, Guevara's dissociation procedure used trypsin, which many researchers avoid due to its poorly understood destructive effects on cell membrane proteins; but it is believed to leave the fast sodium channels relatively intact, whereas van Meerwijk's collagenase procedure may have removed them [Colizza et al. 1983]. This might account for the less abrupt character of van Meerwijk's membrane electrical responses, more like sinoatrial nodal pacemakers than like Purkinje fibers.

Undamaged embryonic pacemaker balls differ importantly from sinoatrial nodal pacemakers in that they use the squid-like mechanism of fast-changing conductivity: the fast sodium channels. So do the tissues into which these cells would have matured, including the Purkinje fibers of the adult heart. If such fibers are removed from the heart, they may

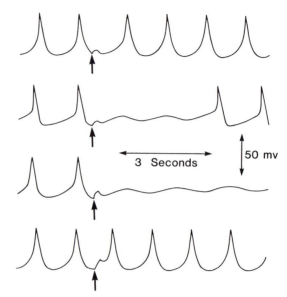

Figure 4.11: A Purkinje fiber fires rhythmically until perturbed by a subthreshold depolarization at the right moment; the action potentials cease, leaving only subthreshold oscillations. Adapted from Jalife and Antzelevitch 1980.

or may not beat spontaneously. In Figure 4.11 (from Jalife and Antzelevitch 1980), a carefully adjusted biasing current keeps a strip of dog Purkinje fiber slightly depolarized (perhaps as in diseased tissue) to ensure regular rhythmicity. This is not the usual cycle of a fully repolarizing, healthy Purkinje fiber, but a cycle based on ionic mechanisms more like those in the sinoatrial pacemaker in somewhat depolarized membrane. Meanwhile its phase resetting is probed by a depolarizing current pulse of carefully adjusted magnitude, applied at various times in the cycle. A critical moment is found when the stimulus eliminates further action potentials, leaving only subthreshold oscillations that sometimes die out or sometimes grow until spiking resumes with an unpredictable phase shift.

Comparable behavior was found in dog ventricular muscle [Ferrier and Rosenthal 1980, even resetting; Rosenthal and Ferrier 1983, annihilation]. As in both the squid giant axon experiment and the Purkinje fiber experiment, phase-resetting behavior was not thoroughly explored: the phase singularity that must envelop any black hole was not exhibited. Neither was a search made for the effective upper and lower limits of stimulus intensity. But the practically interesting point has been made: cardiac pacemakers of diverse origins can be rendered arrhythmic by a critically timed stimulus of the right kind.

Dogs, cats, rabbits, chicks, leeches—all their hearts behave as predicted in respect to this "hit and run" induced arrhythmia. What, then, of a human heart?

At present writing the experiment has been tried only cautiously, in the course of trying to cure unwanted rapid beating of the human ventricle. Such extraneous rhythms sometimes arise in a bit of diseased muscle. Surgical removal of the "ectopic focus," as it is called, incidentally provides a specimen that, if handled carefully, remains electrically active, still beats, and still responds to extraneous stimuli. Robert Gilmour and colleagues in the Krannert Institute of Cardiology in Indianapolis seized the opportunity to test such material for a black hole beyond the phase singularity, and they found it: the ectopic pacemaker in human ventricular muscle could be permanently switched off by procedures not significantly different from Jalife's original test of the theory [Gilmour et al. 1983].

Is it any use to turn off a noxious pacemaker after the tissue is already surgically removed and en route to the incinerator? It might now be worth trying in the living heart, because rapid spontaneous firing (tachycardia) in the ventricle can lead swiftly to fibrillation. The following two experiments do not yet permit definitive interpretation, but it is appropriate to glance at them here because of their potential practical implications. Agustin Castellanos [Castellanos et al. 1984] observed that a normal heartbeat would sometimes switch off an ectopic focus, but only if it chanced to occur at a particular moment in the ectopic pacemaker's cycle. Measuring the phase-resetting curves, they found the critical phase to lie between the large advance and large delay regions—just where the singular annihilating phase would be expected. (Why was it not switched back on by later heartbeats? Unknown, but possibly because of "entrance block" protecting the ectopic focus.) In any case, one cannot rely on chance to control ectopic pacemakers. If they cannot be controlled by drugs, they must be controlled by electrical pulses from an electronic monitoring pacemaker. Can Gilmour's procedure be implemented *in situ* to switch off a tachycardia whenever it crops up?

This challenge was taken up by Rodolphe Ruffy and colleagues at Jewish Hospital in St. Louis [Ruffy et al. 1983]. They introduced the stimulating electrode intravenously, carefully guiding it by hand while fluoroscopically watching its progress into the patient's heart. When the electrode tip arrived at the putative origin of the tachycardia, low intensity pulses were delivered at random times until a 25-millisecond window was defined in which the unpropagated pulse would terminate the ectopic pacemaker's firing. (The stimulus fails to propagate because this window lies in the effective refractory period of the bulk of surrounding ventricular muscle.) Phase behavior was not monitored. Why the pulse size was not reported to be critical and why the pacemaker remained turned off ("entrance block" again?) are also not clear. It might be that these two experiments achieve only the clinical appearance of annihilation, perhaps by stably phasing the ectopic pacemaker to fire in synchrony with the independently triggered ventricles. It might also be that Ruffy et al. electronically annihilated a nearby reentrant wave [Ruffy, personal communication; see also Chapters 5–7 below]. The mechanism of ostensible annihilation is still unresolved, but the provisional interpretations given by Ruffy et al. [1983] refer to the theory of this chapter.

These diverse illustrations stem from a topologically motivated prediction [Winfree 1977] similar to the one that led to the discovery of arrhythmia in circadian clocks. As in that area of physiology, the arrhythmias thus far encountered in diverse living systems exhibit diverse interpretations of "phaselessness" following the singular stimulus. Phaselessness can consist of mere unpredictability, as might be expected if the singularity constitutes only a vanishingly small and elusive target. Or it can fence in a permanent arrhythmia, as when the singularity is stretched thin around an internal black hole. The size and nature of the hueless rim and its potentially hueless interior can only be understood through a properly quantitative biophysical theory; but specialists still hotly dispute the proper choice of equations. Working with the first widely accepted version of the membrane equations for Purkinje fibers [McAllister et al. 1975], T. Chay and Y. S. Lee repeated Best's calculations [Chay and Lee 1984]. Like Best, they used a current bias to stabilize the resting potential as an alternative to self-sustained rhythmicity and found substantial black holes in the contour map of latency. Working next with current-biased ventricular muscle (the Beeler and Reuter 1977 model, which perhaps needs revision of its sodium channel mechanism per Ebihara and Johnson [1980]), they again came up with full-dimensional black holes [Chay and Lee 1985]—as did Jalife, Gilmour, Castellanos, Rosenthal, and Ruffy, working experimentally with diverse parts of mammal and human hearts. In contrast, Peterson's leech hearts, van Meerwijk's and Guevara's pacemaker balls, and Reiner's sinus node model all seem to have only a one-dimensional slit or a zero-dimensional point singularity, providing no resting place for the weary heart. In uncontrived situations there is little reason to expect anything more than the mere mathematical point guaranteed by the coloring theorem: a center of ambiguous latency trapped within an encircling ring of color. In that case, "annihilation" is only as prolonged as the annihilating stimulus is precise, and even this depends on a noise-free physiological environment. In general, rhythmicity can be expected to resume, but at an indeterminate time—as it usually does.

The Whole Heart Is More Complicated

The complexities of timing in the human heart transcend by far the simple-minded descriptions developed up to this point. All our descrip-

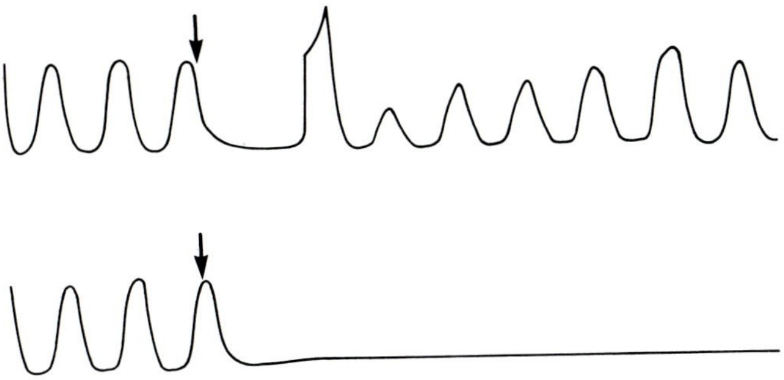

Figure 4.12: In the upper trace, rhythmic contractions in the ventricle of a dog resume after almost any electrical stimulus that avoids the vulnerable phase (which got its name from the publication of these 1939 experiments); but a delicate adjustment of timing annihilates the rhythm in the lower trace, as in Figure 3.1. Though gross behavior of the whole heart here produces a picture similar to Figures 4.6, 4.7, 4.8, and 4.11, single cells here are not quiescent. In fact they are frenziedly active, but no longer in a synchronous way. From Wiggers and Wegria 1939.

tions implicitly ignore the physical dimensions of the heart, as though it were a mere point or a single cell or spatially uniform. But not even a heart as small as a leech's is this uniform. The mammalian heart's pacemaker, the sinoatrial node, is a region composed of many tens of thousands of cells, each coupled to neighbors but capable of beating independently. A whole heart is a multicomponent machine composed of distinctive parts connected by propagation delays. Even if a single fiber or pacemaker center of a real heart were tricked into our condition of indeterminate timing, how long would that condition last in the single fiber, surrounded as it is by others that will fire with some predictable latency?

The question may seem only rhetorical: actively firing neighbors would promptly compel their indecisive neighbor back into decisive cooperation—unless—and here we discover that the question does require an answer—the process that made one fiber indecisive also reset its neighbors to *all* phases of the cycle in a circular gradient of timing around the indeterminately reset point. Chapter 5 will argue that this may be a common situation, and fulfill our foreboding that sooner or later we must take physical space into consideration as an essential ingredient in the timing mechanisms of the heartbeat.

The sinoatrial node (the collection of interacting pacemaker cells considered as a harmoniously functioning unit) is an oscillator and does have even resetting, and it therefore must be liable to some kind of arrhythmia in response to a smaller stimulus, rightly timed. In a spatially structured oscillator, this will presumably be a spatially structured arrhythmia, quite likely involving waves of depolarization and repolarization. As will be seen in Chapter 7, those waves circulate around the singularity.

95

But what about the rest of the heart? Most of the fibers of the adult heart are not spontaneously rhythmic: their rhythm is merely borrowed from the dominant pacemaker. Though it doesn't make a lot of sense to think of them as oscillators in isolation, they *are* oscillating at the last moment of normal timing. And some kinds of tachycardia *are* triggered by a single rightly timed shock, if it is not too small or too big, and do lead to utter arrhythmia (Figure 4.12). The connection with conventional theory of spatially uniform spontaneous oscillators may prove to be subtle, but it is hard to believe there is no connection. A good detective would not give up the investigation quite yet. We have developed only one clue from a physiologist who was following this trail in 1914. A second clue was dropped nearby.

PART II

Rotating Waves

CHAPTER 5

A Clue Involving Space

It is of the highest importance in the art of detection
to be able to recognize out of a number of facts,
which are incidental and which vital. Otherwise your
energy and attention must be dissipated instead of
being concentrated.
—Sherlock Holmes, in
"The Adventures of the Reigate Squires"

An oscillator's phase singularity should not be confused with its steady-
state (alias equilibrium). That mere point in the however-many-dimen-
sional state-space of the oscillator is phaseless, true, since oscillation never
resumes or resumes only at random phase due to random fluctuations
near equilibrium. But it need not be a phase *singularity*, a state surrounded
by neighboring states destined to all phases of the cycle. Even if it is part
of the phase singularity, it is generally an insignificantly small part. The
plainest proof of this insignificance is that in a complicated oscillator it
is usually impossible to achieve a steady state without independently
adjusting all (however many: two or more) variables simultaneously;
whereas a phase singularity, if accessible at all, can always be reached
by adjusting no more than two: the stimulus timing and size (see Appendix).

   The singularity itself is a locus of states near which timing is infinitely
delicate; and if the singularity bounds a black hole, a stimulus on the
black side of the singular locus is guaranteed to lead to no recovery. But
"no recovery" may imply approach to a steady state or it may forebode
transition to a new mode of activity, possibly aperiodic, possibly periodic
in a new way, possibly spatially structured in a new way. This last
"possibility" is a vivid reality in spatially structured oscillators such as
those encountered in diverse chemical and biochemical media (see Chapter 7), in the sinoatrial node, and possibly in the periodically driven
oscillatory parts of the heart, like the ventricular muscle.

Singular Arrhythmia Can Be Spatially Structured

Chapter 3 introduced "A Clue Involving Time." This evolved from Mines's historic triggering of ventricular arrhythmia by a well-timed stimulus of otherwise innocuous size applied to the ventricle during its driven periodic activity. It is also suggested by the clinical observation that periods of tachycardia and fibrillation commonly are preceded by interjection of a premature beat [Katz and Pick 1956, 283-284]. In both animal and human experiments fibrillation is most easily triggered this way during the vulnerable phase of the heartbeat, the electrocardiogram's T wave. An ectopic impulse at this moment can initiate tachycardia (the "R on T phenomenon" [Palmer 1962; Hinkel et al. 1977; Lahiri et al. 1979; Denes et al. 1981; Kempf and Josephson 1984; Geuze and Koster 1984; Hohnloser et al. 1984]). That critical time happens to be near the end of refractoriness, when many fibers have completely recovered from the previous contraction and others have not quite yet recovered: i.e., the moment of greatest spatial nonuniformity, hinting at spatial structure. But not all R on T events trigger arrhythmias. Critical timing is not sufficient; something else must also be critical, possibly the location or the strength of the stimulus. Some such second factor is suggested by the fact that the T wave (and vulnerable phase) occurs near the moments when, in the colored-latency diagram of rhythmic ventricular muscle [Chay and Lee 1985] and of Purkinje fibers [Chay and Lee 1984] the depolarizing phase singularity fleetingly looms overhead *at a critical stimulus strength* (see Chapter 6, Figure 6.8).

In Chapters 3 and 4 we found a way to understand arrhythmias in a formerly rhythmic spatially uniform membrane without having to wait for a clarification of mechanisms: arrhythmia is implicit in even-type phase resetting, in the electrophysiologists's equation for the squid giant axon, in the squid axon itself, in various kinds of heart fibers and in their corresponding ionic equations, and perhaps even in the human ventricle. But in striving to simplify, we neglected to take note of the most conspicuous facts about a real heart: it has spatial dimensions and its major components are not separately rhythmic.

Chapter 6 will investigate what would develop if a spatially extended (not uniform or little) oscillator were shocked with the singular stimulus, implanting a point of singularity somewhere inside. Would the result resemble spatially structured dysrhythmias that precede fibrillation? What are the patterns empirically known or hinted at in typical spatially structured arrhythmias? For this additional clue involving space, we return again to Mines's inquiries of 1914 into fibrillation.

The two clues that preoccupied Mines have developed into two mighty rivers of research that have often seemed so distinct that one of them must be leading off in the wrong direction. In my opinion, they may be different approaches to a unifying synthesis.

The first is the tradition that sees fibrillation as a disorder of impulse initiation. The idea is that, for any of many possible reasons, some fibers become so irritated that they become ectopic foci, firing action potentials spontaneously at a very high rate. If more than one wavefront coexist at all times on the epicardium, then the electrocardiograph trace will be confused and may be called "fibrillation." Ideker et al. [1981] show exactly this situation in a startlingly clear and decisive epicardial mapping experiment. How can such disorders of impulse initiation arise? At the borders of an infarct—a region of cell death where a plumbing blockage, a coronary occlusion, has cut off the supply of oxygen-bearing blood—there is a great deal of abnormal electrical activity. More commonly, a temporary local occlusion (Prinzmetal's angina or the less conspicuous "transient myocardial ischemia," both due to spasm in a coronary artery) temporarily makes the muscle more susceptible to arrhythmias, in part by eliciting neural traffic from the brain [Shepherd 1985; Skinner 1985]. Or one of the many nervous inputs to the heart might cause an untimely local irregularity, as they commonly do when organs are handled during abdominal surgery [Bellet 1971, 585, 845-849]. In any case, Mines had the idea that the disorder of impulse initiation might be ephemeral; the original provocation might not need to persist once it had triggered tachycardia. He therefore studied the effects of a randomly timed little jolt. This provided our first clue, the one involving time in Chapter 3.

The other clue stands at the source of the other river of research, in which fibrillation—or at least its maintenance once initiated—is seen as a disorder of impulse conduction. A fibrillating heart, continually alive with fine-grained rapid activity, never rests until exhausted; but need we impute all this activity to point-like irritable foci? Maybe fibrillation gets its start from normal impulses propagating at intervals too short for the medium to respond reliably, resulting in local fragmentation of wave-fronts and even creation of vortex pairs [Allessie et al. 1976; Boineau et al. 1980a; Pertsov et al. 1983; Zykov 1984; El-Sherif 1985; El-Sherif et al. 1985; see pp. 114ff. below]. The common factor, regardless of the initiating process may be circulation of the impulse. Just before the First World War, Mines in Montreal and W. E. Garrey in San Francisco were studying the stability of impulses circulating on doughnuts of muscle [Mines 1914; Garrey 1914]. Their work, much amplified by generations of ingenious experimenters and theorists, provides our second clue, the one involving space.

A Breakdown of the Arrangement of Timing in Space

To develop this second clue, we begin with the recognition that fibrillation in healthy muscle is initially (before coronary circulation falters and fibers fatigue) an organizational disease. The undamaged single fiber does not "fibrillate." Its behavior during fibrillation is a normal response to the irregular stimuli impinging on it from other normal fibers (except that they all become somewhat depressed during high-frequency activity, of course, and except in the ischemic tissue that commonly gets embroiled in fibrillation). Garrey and Mines realized this and concluded that fibrillation has to be understood primarily in terms of an altered spatial arrangement of conduction. Their feeling was based on three observations.

First, fibrillation is a spatially structured condition. The fibrillating muscle is not uniform or synchronous in its activity. In fact, it was later found that the best way to cure fibrillation quickly is to impose spatially uniform synchrony, e.g. by massive electroshock or widespread infusion of potassium salts [Zipes 1975].

Second, fibrillation is easier to trigger in tissue that is intrinsically nonuniform or has been made so at least temporarily, for example, by a regional failure of circulation, by the recent arrival of a spatially graded barrage of impulses from the vagus nerve or a sympathetic ganglion, by 60-Hz electric shock, or by anything that decreases the speed of conduction (so that a greater range of phases is present simultaneously). Much emphasis was placed on the importance of spatially graded stimulation and local islands of conduction block as essential preconditions for the onset of the tachycardias that degenerate to fibrillation.

Third, fibrillation is stable only in big hearts. In little hearts or in big fibrillating hearts progressively trimmed with a paring knife, fibrillation is unstable. Moreover, fibrillation has a fundamentally two-dimensional or three-dimensional local structure: it cannot be propagated through an isthmus of tissue narrower than about 1 cm, even though individual impulses propagate through without difficulty [Garrey 1924].

In any case, Mines and Garrey both speculated that fibrillation might be an extreme development of milder dysrhythmias whose essential feature is circulation of an impulse: what physiologists today call reentry [Wit and Cranefield 1978; Wit and Rosen 1981]. Let me explain how they viewed it.

Electrical Waves in Heart Muscle

Electrical impulses in nerve membrane or heart muscle propagate without attenuation, unlike sound waves or ripples from a raindrop striking water (but like the chemical waves that will be discussed in Chapter 7). The signal from sinoatrial pacemaker loses no strength as it spreads. Each little fiber of heart muscle contributes to the wave as it goes by, restoring it to full strength as in a Pony Express relay. As it was with Pony Express, so it is with the human heart: each station needs a little while to recover before it can again contribute to a passing signal. In the heart this interval is called the refractory period.

Normally the ventricular chambers of the heart are activated everywhere simultaneously along their inner surfaces. Except in the interventricular septum and part of the right ventricular free wall [Durrer et al. 1970; Brusca and Rosettani 1973], excitation spreads through the muscle mostly from inside to outside in parallel, without propagating extensively in the transverse direction. As Garrey put it [Garrey 1914], "the ventricles beat apparently as a unit but in reality as a group of isolated segments." But if the signal should somehow be deflected to propagate transversely, what patterns might prove stable? In particular, if it could propagate in a circular path, it might circulate forever—but only if the ring is big enough (and propagation slow enough) to allow recovery before the wave returns. It seemed to the early physiologists that waves might circulate perpetually around a big enough ring of heart tissue. To find out, they made such rings: first out of jellyfish umbrella, then out of sea turtle heart or dog heart. The jellyfish proved able to sustain a circulating impulse for six days (something like half a million cycles, close enough to "perpetually" [Mayer 1906]). By cutting an orifice in the middle of a cookie-size piece of ventricle (Figure 5.1) they learned that, rightly stimulated, heart muscle also does indeed support an undamped circulating impulse for at least some thousands of circuits [Mines 1914; Garrey 1914; Garrey 1924].

Propagation around a loop causes one version of a clinical dysrhythmia called the Wolff-Parkinson-White syndrome: a cure is often effected by surgically severing some accessory pathway that completed a loop between atrium and ventricle. Another dysrhythmia called atrial flutter is commonly believed to involve a wave circulating around the orifices through which the major plumbing enters the atria. In the ventricles such a wave may also circulate around a patch of scar tissue, the legacy of a coronary obstruction. In any case, if the rotation period is less than the sinoatrial pacemaker's period, the circulating pulse may usurp the pacemaker function and we have a seriously persistent dysrhythmia to contend with.

This is one mode of reentry: circulation on a physical ring of tissue. This traditional and usual meaning of the term "reentry" connotes an essentially one-dimensional source: a pulse circulating on a loop of one-dimensional pathways. It can be initiated by a stimulus that momentarily fails to propagate in one of the two directions along the ring. The pulse

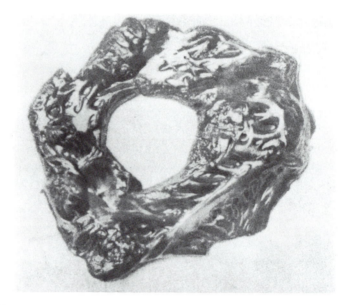

Figure 5.1: A ring of ventricular muscle about 6 cm in diameter, cut from the heart of a dog. From Mines 1914.

circulates with a period proportional to the size of the ring. It can be eliminated by a single stimulus placed just in its wake.

But there is another, quite distinct mode that may be the more common prelude to fibrillation: a mobile vortex-like rotation of a wave in healthy tissue, requiring no loop or hole to define a circular path. It is sometimes called "micro-reentry" to distinguish it from the long loops commonly involved in "macro-reentry." This second mode can only arise in a medium of at least two dimensions. In two dimensions it resembles a vortex; in three, a smoke ring. It is initiated in amazingly diverse situations. Unlike classical one-dimensional reentry, it can be eliminated only by stimulation of a region big enough to span counter-rotating vortices or to reach to the boundaries of the medium. It can (and does) migrate freely through the medium unless it should chance to encounter a hole and remain snagged to circulate around it. Consisting of an extended wavefront, not just a pulse, it never really "reenters" a previously excited region, but some other part of the same wavefront does. Perhaps we should not use the traditional term "reentry" at all in this connection; but the usage has already begun, so it is continued here, often with a distinguishing adjective. This vortex-like reentry is the mode revealed in multielectrode mapping observations of some tachycardias such as flutter and, in some cases, early stable stages of fibrillation [Allessie et al. 1985]. (The majority of clinically important arrhythmias are undoubtedly much more complicated.)

Rotating Waves

At the turn of the century, physiologists still wondered whether such a wave could stably circulate in unperforated tissue such as healthy ventricle. Because it was not feasible then to answer the question by direct observation, they turned to mathematical caricatures of nerve and heart membrane called excitable media. In the simplest excitable medium (see Box 5.A below) each cell may be either excited, refractory (for a brief time after excited), or quiescent. In the latter state, it will turn excited a moment after any immediate neighbor turns excited. Thus excitation propagates. Can a block of unperforated excitable medium support vortex-like circulating activity? A clear affirmative answer was first given by experiments on turtle heart: circuits as short as 3 to 6 cm were observed in muscle without holes ("natural rings of tissue are not essential" [Garrey 1914; Garrey 1924]) and these vortices were conceptually linked to tachycardia and fibrillation. But later theorists disagreed: subject to inevitable simplifications that seemed innocuous at the time, such a wave was convincingly argued to be impossible. The experiments were apparently discredited.

This verdict came with no less authority than that of Norbert Wiener, the mathematical genius who worked with Mexican cardiologist Arturo Rosenblueth on this conundrum in the middle 1940s [Wiener and Rosenblueth 1946]. Their model was promptly improved somewhat and found to admit a rotating wave, but still, under the approximations then entertained, the rotating wave seemed unstable [Selfridge 1948].

Wiener and Rosenblueth adopted only one of many alternative ways to simplify reality. They assumed no harm would come of neglecting the finite duration of the action potential in heart muscle or replacing electrotonic coupling between adjacent cells by a spatially and temporally discontinuous rule. Another caricature [Katz 1946] entailed a similar vision of excitability but supposed less orderly circulation: essentially a free-for-all of wandering wavelets, seldom returning along prior paths. This notion was developed by cardiac physiologists Gordon K. Moe and J. A. Abildskov at S.U.N.Y. Upstate Medical Center in Syracuse in the early 1960s [Moe and Abildskov 1959; Moe 1962]. They reasoned that if they could not only discretize the representation of (five) possible electrical states of the fiber but also coarsely discretize the tissue as well (by presupposing complete electrical uniformity within each 4-mm by 4-mm

Box 5.A: A Cellular Excitable Medium

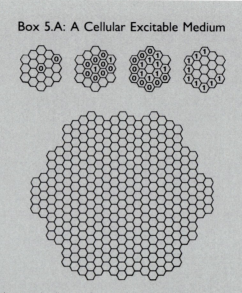

Figure 5.2: Hexagonal cell array for manual simulation of excitable media. The top row provides initial conditions and three further steps by application of the rules.

Take a pencil and a sheet of tracing paper and play with Figure 5.2 according to the following game rules: (This game is a lot like Conway's Game of Life [Gardner 1970, 1971]. It has been elaborated a great deal in the context of biological excitable media by Reshodko [1973], Reshodko and Bures [1975], Greenberg et al. [1978], and Hastings [1981] and in the context of cosmology by Madore and Freedman [1983].)

Each little hexagon in this honeycomb is supposed to represent a cell that may be *excited* for the duration of one step (put a "0" in the cell) or *refractory* (after the excited moment, replace the "0" by a "1") or *quiescent* (after that erase the "1") until such time as any adjacent cell becomes excited: then pencil in a "0" in the next step.

If you start with no "0's," you'll never get any, and this simulation will cost you little effort. If you start with a single "0" somewhere, it will next turn to "1" while a ring of 6 neighbors become infected with "0". As the hexagonal ring of "0's" propagates, it is followed by a concentric ring of "1" refractoriness, right to the edge of the honeycomb, where all vanish.

Now see what happens if you violate the rules just once by erasing a segment of that ring wave when it is about halfway to the edges: you will have created a pair of counter-rotating vortices (alias phase singularities), each of which turns out to be a source of radially propagating waves.

(Stop reading until you have played some.)

You may feel a bit foolish, since this is obviously supposed to mimic action potential propagation, and the caricature is embarrassingly crude. Which aspects of its behavior are realistic and which others are merely telling us "honeycombs are not heart muscle"? The way to find out is to undertake successively more refined caricatures until a point of diminishing returns is reached. For most purposes, it is reached surprisingly soon.

The biggest single step forward was also the first, a refinement of these game rules to incorporate nonuniformity in the duration of refractoriness and other intrinsic parameters of the cell membrane. In 1964, Moe, Rheinboldt, and Abildskov [Moe et al. 1964] prevailed upon a digital computer to do the bookkeeping on a large scale for their quantitatively more cumbersome rules. This might seem unimpressive in a day of video terminals and home personal computers, but their printouts of wave movements from millisecond to millisecond were a conceptual tour-de-force then, and they revolutionized thinking about fibrillation. A more realistic study along the same lines was conducted a decade later by Foy [1974], and another by Smith and Cohen [1984].

The results, in short: Wavelets could meander along ever-changing tortuous paths without remission, much as had been imagined by Garrey [1924], Wiener and Rosenblueth [1946], and Katz [1946]. This condition (two-dimensional computer "fibrillation") required a certain minimum of space and then could be triggered by a premature impulse arriving in the midst of inhomogeneous tissue. Moe et al. [1964] found that "[T]he induction of fibrillation should be facilitated by any agency which increases temporal dispersion of excitation and recovery." The main condition to start fibrillation was a "dephasing of closely adjacent units." This is a fair summary of the essence of a phase singularity.

Reentry was instigated in these first simulations of excitable media through the mediation of irregularities (viz. inhomogeneous recovery rates). This drew much attention to the role of nonuniformities. Beautifully clear experimental illustrations are provided in the publications of Allessie, of El-Sherif, and of Boineau; mechanisms are clarified computationally in many papers from the Puschino/Moscow biophysics community [e.g. Pertsov et al. 1984; Zykov 1984, Figure 5.6]. But it is important to remember that inhomogeneity is not a necessary catalyst. Two-dimensional reentry (rotors) can also arise in spatially uniform but anisotropic tissues due to their diminished conduction speed transverse to the "grain" of the muscle [Spach and Kootsey 1983] and just as easily in uniform and isotropic tissues due to gradients of local stimulation [Gul'ko and Petrov 1972; Winfree 1974b; Winfree 1978]. As van Capelle and Durrer [1980] concluded: "Dispersion in the intrinsic properties of the individual elements, however important its role, is definitely not necessary for this type of arrhythmia." As the more complex situations dominate the literature, my effort here is to emphasize the simpler underlying principle.

block of two-dimensional muscle), then they could cajole a computer into simulating the whole affair in a reasonable time and not have to worry about intractable mathematics.

Moe and Abildskov collaborated with computer programmer W. C. Rheinboldt to simulate successfully something very much like Garrey's idea of multiple wandering wavelets: the computer produced two-dimensional fibrillation in a discrete network of cardiac fibers variegated by random assignment of refractory periods [Moe et al. 1964] (Box 5.A). When they made all the 4-mm blocks identical in electrical properties, their "fibrillation" simplified to a pattern reminiscent of the circulating excitations now known to precede fibrillation: the turbulent activity resolved itself into an assortment of stable vortices.

Moe et al. had clearly computed the "forbidden" rotating vortices. So had B. G. Farley [1965] during his computer simulations of randomly connected nerve networks: "There are many different modes of self-excited oscillation possible. A particularly pervasive and striking one consists of one or more continuously rotating spirals." But Farley apparently overlooked their possible role in cardiac arrhythmias.

Even this could still appear as "fibrillation" to an electrocardiogram. Sixty years ago T. Lewis [1925] already recognized that the irregularity of the EKG in fibrillation is frequently more apparent than real, often appearing irregular only transiently in one lead while another lead shows quite regular short-period oscillations. Electrograms taken locally at the surface of the heart are indeed often regular and discrete during fibrillation [Josephson et al. 1979], as are more remote electrocardiograms [Nygards and Hulting 1977; Kuo and Dillman 1978; Herbschleb et al. 1979; Tabak et al. 1980; Herbschleb et al. 1980; Herbschleb et al. 1982; Forster and Weaver 1982]. Ideker et al. [1984] also emphasize that it is difficult to see how the EKG can show its common strong signal in the 10–12 Hz band if fibrillation really consists of many randomly wandering wavelets: whatever fibrillation may be, it is better organized than that, at least in the early stages.

Thus the Wiener and Rosenblueth model now seems an extreme limiting approximation for badly inhomogeneous tissue; fibrillation and prefibrillatory tachycardias in more uniform tissue may instead consist of two-dimensional vortices. It would be of interest therefore to study their properties.

About the time of Moe's computational experiment with a deliberately discretized medium, Russian mathematician I. S. Balakhovsky began to reexamine the problem from yet another perspective. Balakhovsky [1965] retained continuity of the medium and additionally made Wiener's discrete-state caricature a little more realistic for heart cells. He analyzed circulation around a hole and was able to sew the hole shut and still obtain vortices stably rotating about a slit-like arc of functional conduction block. But would such imaginary waves be stable in the face of minor variations of refractory period, excitability, etc., that must be reckoned with in living tissues? And would they remain viable when it comes time to admit that real tissue has a continuum of states intermediate between the three discrete states of this caricature?

Biophysicist V. I. Krinsky, working in the Institute of Biological Physics of the USSR Academy of Sciences near Moscow, refined the physiological assumptions still more by making both the medium and the electrical state continuous. Krinsky showed mathematically how a wave—which he dubbed "reverberator"—could rotate stably, both faster and in a much smaller space than had been imagined [Krinsky 1966 and Krinsky 1968 in Russian; Krinsky 1978]. During the next decade Ph. Gul'ko, A. A. Petrov [Gul'ko and Petrov 1972], A. I. Shcherbunov and colleagues [Shcherbunov et al. 1973], J. L. Foy [Foy 1974] and myself [Winfree 1974ab] and F. van Capelle and D. Durrer [van Capelle and Durrer 1980; see Figure 5.3 below] published computerized caricatures of wave propagation in excitable media like heart muscle. All revealed rotating waves with varying degrees of stability. In short, after seventy years we have come back to the observations and ideas first proffered by Garrey and Mines; the early theorists Wiener and Rosenblueth were simply wrong in their conclusion that waves could rotate only around a physiological hole of substantial dimensions. (The mathematical errors were first isolated in publications by Selfridge [1948], Abakumov et al. [1970], and Durston [1973].) When diffusion of electric potential is realistically taken into account in the next generation of models after Krinsky's, one finds in place of the hole, at the source of Krinsky's spiral reverberator, a tiny rotating generator called a "rotor" [Winfree 1978].

Complications

This second mode—the wave rotating freely, not channeled in a specific loop of muscle or constrained to circulate around a specific lesion—can be quite mobile, like a wandering tornado. First of all, the rotor typically drifts transversely to gradients in the medium's properties [Rudenko and Panfilov 1983]. But even in a uniform medium the rotor may wander, as in Figure 5.3 (see also Box 7.B). It is not necessary to invoke irregularities in the medium (though they may exist) to account for erratic behavior of this labile vortex. The physical conditions that distinguish unstationary rotation from stationary rotation on a uniform medium without holes have been explored by Zykov [1984]. In a nutshell, if the

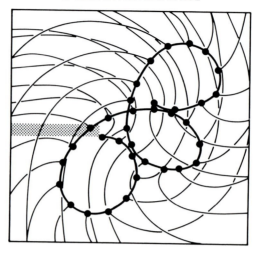

Figure 5.3: By stimulating the stippled region just behind the refractory wake of a plane wave (not shown, just exiting the bottom of the square) a hemi-wave is created that circulates about a moving pivot for at least 50 cycles. The wavefront is best followed by noting the successive positions of its endpoint, marked by filled dots along the heavy path that begins at the end of the stippled rectangle. Trace it through the first three loops of the wavefront's nominal endpoint. Adapted from van Capelle and Durrer 1980, Figure 10, a numerical caricature of 529 coupled cells intended to mimic a completely uniform 13-mm square of depressed cardiac muscle. Their underlying 23 × 23 grid is rather coarse for this purpose and the simulated slow-response kinetics is rather arbitrary, but more refined calculations behave similarly in respect to rotation and meander.

initial response to triggering is too quick or the action potential duration is too long (or both), stationary rotation becomes unstable and meander begins.

Worse, such a wave can multiply. Like Moe and Abildskov in 1964, Krinsky and his colleagues found in their more realistic simulations that single rotating waves can fragment, multiply, and spread if the tissue contains nonuniformities such as commonly result from ephemeral local obstruction of the coronary blood supply. In fact there are sufficient nonuniformities even in the healthiest heart. Wiener and Rosenblueth expressed this in their 1946 classic on rotating waves:

> [I]n the case of fibrillation we cannot ignore the heterogeneity of the muscle, if fibrillation is a random, small scale series of wavelets. Under these conditions the specific connections between neighboring fibers become more important than the way in which the tissue made out of these fibers approximates to a homogeneous structure.

Their impression is copiously confirmed in the subsequent literature of cardiac physiology. Cardiac muscle is fibrous and directional not only in its gross anatomy and mechanical behavior but in its electrical properties as well. The laboratory experiments and computer simulations of

M. S. Spach at Duke University illuminate the circumstances in which this anisotropy plays a dominant role [Spach et al. 1982; Spach and Kootsey 1983]. The innervation of the heart is even more irregular [Alessi et al. 1958; Han et al. 1964; Han and Moe 1964; Ninomiya 1966; Armour and Sinha 1975; Kralios et al. 1975; Wit et al. 1975; Randall 1977; Armour et al. 1972; Martins and Zipes 1980]. How much does all this realism spoil the simple picture of waves serenely rotating in an idealized excitable medium? In one limiting approximation, it is totally disruptive—e.g. see Smith and Cohen 1984. No one really knows for sure where real arrhythmias lie between the extremes of perfect coherence in the normal heartbeat and utter chaos in terminal fibrillation (Box 5.B). The notion explored here and in Chapter 6 is that irregularities probably help in the creation of rotors and probably modify their behavior once established, but not so fundamentally as to make the basic notion useless. Accordingly this chapter and the next emphasize only the central idea in stark simplicity. In Chapter 7 the same concepts will be carried into perfectly isotropic and uniform *chemical* media.

While our central purpose is to keep the simple ideas in the foreground, this is not a book just about ideas. It is now time to revert again to experiment, for, as John Wheeler once remarked in another context [1964], "It is too much to imagine that one has yet made enough mistakes in this domain of thought to explore such ideas with any degree of good judgment."

The First Clear Glimpses of the Rotor in Heart Muscle

In both living and nonliving experimental systems to be discussed in Chapter 7, one can actually see pirouetting patterns of electrochemical activity with the naked eye. But in heart muscle they are far too fast for unaided vision: one rotation takes only a fifth of a second or less. Moreover, the muscular contraction set off by the passing wave is faint: the muscle never gets a chance to relax fully before the next wave is upon it. Physiologists have therefore resorted to electrical recordings. A grid of many separate electrodes is placed on the surface of the heart, each electrode backed by an amplifier to pick up the faint signal and store it. The saved signals must then be processed to determine when waves pass each point. The points passed simultaneously are connected: that is a wavefront; and the wavefront is located anew every 10 milliseconds or so. Its successive positions are called isochrons or isochronal contours [Moe et al. 1941]. If they pirouette around some central region of ambiguity, rotating like windmill blades, then we are seeing a rotor.

L. Rozenshtraukh and A. V. Kholopov in the Soviet Union made the first published serious effort to demonstrate such a rotor in the heart, initiating it by localized vagal discharge in the atrial muscle of frog heart

Box 5.B: Different Ideas

Gordon Moe and colleagues developed the first clear articulation of ideas about the mechanisms underlying fibrillation. Their computer simulation to test the implications of their idea was far ahead of its time [1964] and still guides both experiment and theory in this area [Smith and Cohen 1984; Allessie et al. 1985]. They emphasized the nonuniformity of electrical properties to derive meandering wavelets in a two-dimensional, discontinuously cellular, discrete-state vision of fibrillation. Their result captured much of the spatial structure of fibrillation, but it left temporal structure unaccounted for and in fact denied that there should be any.

Twenty years later, there are of course additional data and the corresponding additional speculations weaving together spatial and temporal patterns. One such speculation is presented in this chapter and the next. Another alternative abandons spatial structure completely to account for the observed temporal regularities of fibrillation. Johan Herbschleb of the Utrecht Ziekenhuis Cardiology Department argues that in fully developed three-dimensional (ventricular) fibrillation, disorganization proceeds even further than in Moe's wandering wavelet model [Herbschleb et al. 1982; Herbschleb et al. 1983]. Herbschleb noticed that the electrocardiogram is by no means arrhythmic during the first minutes of ventricular fibrillation, even though it does lack the familiar signal complexes of an orderly heartbeat. In fact it is often strikingly regular [Nygards and Hulting 1977; Kuo and Dillman 1978; Herbschleb et al. 1979; Tabak et al. 1980; Herbschleb et al. 1980; Herbschleb et al. 1982]. (The same is true in atrial fibrillation. As early as 1965, Battersby [1965] recognized the challenge that this regularity presents to the wandering wavelets model.) In place of the usual once-a-second P (atrium firing), QRS (ventricle firing), and T (ventricle recovering) deflections (Figure 2.4), one commonly sees a regular rhythm of mixed 5 and 10 Hz components (similar to Figure 2.5). The 5-Hz component is typical of ventricular tachycardias (300 bpm) and of the refractory period of ventricular fibers. The 10-Hz part, more characteristic of fibrillation and not simply a harmonic artifact, taxes the individual fiber's ability to repolarize in time for the next depolarization; could it be that fibers are firing one another alternately? Intracellular microelectrode recordings during fibrillation should suffice to distinguish whether all fibers fire at 10 Hz, or only at 5 Hz in alternation; adequate data are not yet available.

In Herbschleb's model the ventricular muscle is organized into two groups of cells regularly alternating like an antiphonal chorus, but with no vestige remaining of spatial organization. This "choral" mode has been remarked upon in theoretical models of circadian (24-hour) rhythms at least since 1965 (see Winfree 1980, Chapter 11) and in theoretical neurophysiology since 1967 [Krinsky and Kholopov 1967]. Much more complicated versions have been dissected theoretically [Honerkamp 1983; de Bruin et al. 1983; Strittmatter and Honerkamp 1984]. Its independent discovery in present context (supposing Herbschleb's model is confirmed in the laboratory) has new clinical interest because the abrupt onset of this high-frequency coherence may provide the best cue to a microprocessor-controlled defibrillator implanted in the patient's chest [Forster and Weaver 1982]. Reliable cueing is essential: the defibrillator must deliver its resynchronizing jolt without delay if fibrillation is really starting; but if it misfires on ambiguous warning it may start fibrillation!

To nucleate choral antiphony in a population of formerly synchronous nerve cells, one must first scatter the firing times of nearby fibers. This seems to be just what the singular impulse does during the vulnerable phase: it provides the seed of arrhythmia. What it grows into depends on circumstances. Herbschleb envisions no spatial organization whatever, but a spatially structured alternative is presented in Chapter 6. Maybe all current notions are correct, as stages in a progression. When ischemia and nonuniformities are contemplated, scenarios multiply profusely. Perhaps first there is an irritable focus of such high frequency that one activation wave is not finished before a second and third have already left the focus [Ideker et al. 1981; Ideker et al. 1984; Bardy et al. 1983]. Maybe the ostensible ectopic focus is really reentrant in mechanism. Maybe it is the entire web of ischemic Purkinje fibers chaotically firing impulses into the muscle from the endocardial surface. Whatever the source, its short-period output taxes the other fibers to follow. Waves propagate but soon fragment, especially upon encountering islands of slightly-too-lethargic tissue. Their broken edges become paired singularities that turn with somewhat longer period [Allessie et al. 1976; Boineau 1980a; El-Sherif 1985; El-Sherif et al. 1985]. If there are too many such, the fine-grained structural irregularities of heart muscle may reduce all to mere turbulence. In this chaos, the spatially unstructured "choral" alternation nucleates, and finally the muscle perishes of exhaustion.

Clearly, there is still too much room for fantasy in this business and much remains to be learned. My own approach is to first try to understand reentry in uniform healthy membrane, and even before that, to try to understand it in mathematical and chemical models of excitability.

[Rozenshtraukh et al. 1970]. Their findings were not convincing due to technical limitations on the number of electrodes they could deploy simultaneously. In 1972 the first clear observation of rotors in heart muscle resulted from the determined resourcefulness of M. A. Allessie, F.I.M. Bonke, and F.I.G. Schopman in the University of Amsterdam's Physiological Laboratory [Allessie et al. 1973; Allessie et al. 1976; Allessie et al. 1977]. Intrigued by the observation that a single well-placed and well-timed 1-msec pulse only 4 times the diastolic threshold would often evoke repeated rapid discharges, they undertook to examine the spatial pattern of this repetitious self-activation. They systematically moved an array of ten electrodes across a piece of rabbit atrium in which they had initiated this dysrhythmia (atrial flutter) by prematurely stimulating a point just behind a passing wave. They were able to reconstruct the positions of its wavefront as a succession of isochronal contours. The contour lines radiated from a pivot (or a pair of counter-rotating pivots) like the blades of a turbine, revealing a rotor that spun ten times per second, minute after minute. As the wave rotated in this electrical vortex, the inner endpoint of each isochron described a loopy path 6 to 8 mm in diameter. The radius of the loop was thus only a few times more than the space-constant for electrical coupling in this medium. This was somewhat smaller than the minimum mass of 32 mg (about 25-mm perimeter) required for persisting fibrillation, as roughly determined by successive halvings of rabbit atrial muscle [West and Landa 1962]; the rotation frequency, 10 Hz, was about the same.

In the 1976 experiment of Allessie et al. [1976] (Figure 5.4) two mirror-image rotors arose when a stimulated wavefront broke around both sides of an island of still-inexcitable fibers, not yet fully recovered from the immediately prior wave [El-Sherif 1985, Figure 16; El-Sherif et al. 1985, Figure 6].[1] One rotor, being too near the tissue's edge, proved unstable, leaving only its better-situated twin. Instabilities in the initial symmetry were first anticipated in this context by Foy [1974]. In a surprisingly deep analysis of fibrillation and defibrillation (considering that it remained unpublished), Foy simulated two-dimensional (atrial) myocardium numerically, observing that "the present model demonstrates how a reentry can develop [in this case following a localized conduction failure in a region of depressed excitability] and how such reentries naturally lead to pairs of self-sustaining spiral waves; and how, frequently, one of the spiral waves in such a pair is crowded out by its twin or fails in some other way" [Foy 1974].

Frequently, however, is not always. Recent isochronal mapping experiments in ventricular muscle reveal stable pairwise creation in the equivalent but roomier situation of the dog's ventricle [El-Sherif et al. 1982; Wit et al. 1982; Mehra et al. 1983]. We will see in Chapter 8 that they probably go deeper, even in relatively thin atrial myocardium, form-

---

[1] Another interpretation would emphasize that the rotors both arose within a few space-constant's distance from an electrode that provided several-times-threshold excitation in the midst of a region where coupling interval was graded through the vulnerable phase. The possible significance of these coincidences is developed in the next chapter.

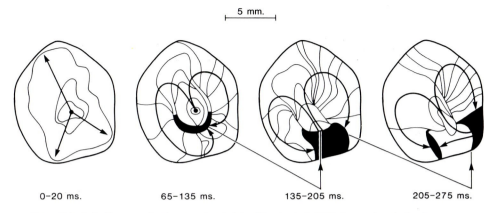

| 0-20 ms. | 65-135 ms. | 135-205 ms. | 205-275 ms. |

Figure 5.4: Activation wavefronts reconstructed by Allessie et al. 1976 from microelectrode re-cordings in the atrium of a rabbit. The contour lines are isochrons (successive positions of the wavefront) at 10-msec intervals. Superimposed arrows run from the beginning to the end of propagation in each panel. The first panel shows one beat of 2/sec paced activation fired from an electrode (black dot). The second panel shows activation from an extra (vulnerable-phase) stimulus 65 msec later, mostly blocked below but escaping upward and around to left and right as paired rotating tachyarrhythmia. Continuation of the wavefront from one panel to the next is indicated by arrows connecting the last isochron in one panel to the first isochron in the next during 70-msec rotation (about 14 Hz). The two rotors are initially about 2 mm apart: each about 1 space-constant removed from the vulnerable-period stimulus. The anticlockwise wave on the left fails at the end of the third panel and again at the end of the fourth panel. The original publications from which this figure was adapted use a discrete color code, not to be confused with the use of continuously graded color throughout this book to represent phase in a cycle. Adapted by per-mission of the authors and the American Heart Association, Inc.

ing three-dimensional vortices that erupt like sunspots in counter-rotating pairs on the surface, or, if unpaired, a solitary tornado.

Returning to Allessie's original demonstration, Zykov [1984, Chapter 8] additionally interprets it as an instance of stationary rotation—as indeed it must be, considering that Allessie's methods at the time were incapable of mapping anything that departed from strict repetitiousness. Less sure is whether the rotor is stationary for the dynamical reasons alleged by Zykov or because it is caught on an island of more refractory tissue in the rabbit atrium.[2]

Using dogs instead of rabbits, with 960-electrode technology a decade later, Allessie et al. 1984 (their Figure 4) show isochrons at 3-mm res-olution, spreading in rings from the sinoatrial node in a normal beat. In

[2] Zykov raises a conceptual issue of potentially greater significance than the nature of this particular example. He finds by numerical simulations that the intuition universal among cardiac physiologists—that wave circulation requires more space if the refractory period is longer and/or wave speed is greater, as Wiener and Rosenblueth first reasoned—is correct only in the case of unstationary rotation. When parameters are such that rotation

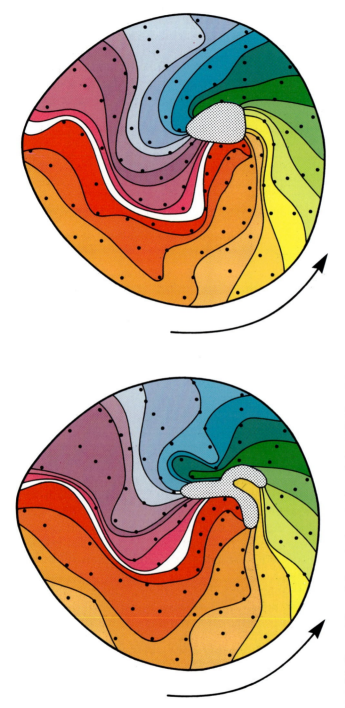

Figure 5.5: Two rotations of the activation wavefront during artificially induced ventricular flutter are indicated by 12-msec isochrons, color-coded on two consecutive (left, then right) polar projections of the entire surface of the human ventricles. In each picture a white "crack" is left behind the first isochron, because the last colored isochron marks the wavefront just before it has returned to the initial position. The pivot is near the apex (bottom) of the heart, in the center of each picture. This region of irregular timing is stippled grey in each. It is not clear whether there is a rotor (source) around the pivot, or only unexcitable diseased tissue (a functional hole). The dots are positions of electrodes at which the passing wavefront was detected electrically for storage and computer display during an open-chest operation. (The pivot area thus identified was cut out minutes later.) Adapted from Downar et al. 1984 with permission from the authors and the American College of Cardiology.

contrast, the same electrodes reveal the pinwheel-like arrangement of isochrons during 420 bpm atrial flutter caused by a meandering rotor.

A single wavefront rotating 4 times per second in the human ventricle (quite likely snagged on a "hole" of depressed excitability in this instance) is depicted by 110 epicardial electrodes in Figure 5.5 [Downar et al. 1984]. This lethal condition is called ventricular flutter. Harumi et al. 1980 found vortex-like reentry of about the same period in the epicardial muscle of a dog's right ventricle; and Richards et al. 1984 observed much the same in the surviving epicardium above an infarct in dog ventricle. In this case they found that the rotor persisted in a piece only 2-mm thick and about 16-mm square: stability in a space scarcely larger than the 6–8 mm diameter rotor found above in the atrium.

These views of prefibrillatory tachycardia from the epicardial (outside) surface of the heart or from the endocardial (inside) surface of extirpated bits of the atrium can now be supplemented by views from the endocardial surface *in situ*. In Allessie's University of Limburg laboratory in Maastricht a bundle of 480 fine electrodes is mounted inside a little dimpled "egg" or "strawberry" (Figure 5.6, top) that fits snugly inside a dog's atrium [Allessie et al. 1982; Allessie et al. 1984; Allessie et al. 1985]. This, of course, prevents the pumping of blood, but circulation can be provided artificially during the experiment. The electrodes contact the entire surface of the atrium at intervals of about 3 mm. For work on the epicardial surface of the ventricles [Wit et al. 1982], a stretchable fabric "sock" or a disk-like placque carries 384 electrodes, spaced 2 mm apart (Figure 5.6, bottom). Each electrode conveys the local electrical rhythm to a computer, whose job it is to sample them all fast enough and keep the data sorted out. Later these voltages are assembled into isochronal maps showing how the pattern of electrical activation changes from moment to moment. Figure 5.7 shows a fairly simple, clean wave circulating around a point—and its complex deterioration into turbulent eddies of fibrillation (see Box 5.C).

What does fibrillation look like in ventricular muscle? Using a sta-

---

is stationary, as in this first experiment, the space required varies with refractory period and wave speed in a way exactly opposite to conventional expectation, making nonsense of the customary criteria for prescribing anti-arrhythmic medications! Qualitative departure of rotor behavior from the "intuitive" expectation first articulated by Wiener and Rosenblueth is not limited to this instance; Mikhailov and Krinsky [1983] and Pertsov et al. [1984], for example, also demonstrate spiral waves with period much in excess of the refractory period, circulating with a large excitable gap between wavefront and the receding refractory wake ahead of it, circulating not around a hole, or even around the rotor core, but around an expanse of medium that remains unexcited only for the reason that the wavefront cannot curve sharply enough to get inside [Zykov 1980; Zykov and Morozova 1979]. This computational phenomenon seems wholly new and might reasonably be sought in cardiac muscle.

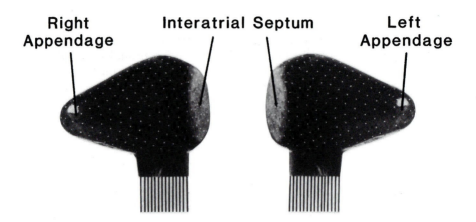

Right Appendage **Interatrial Septum** **Left Appendage**

2 x 480 leads

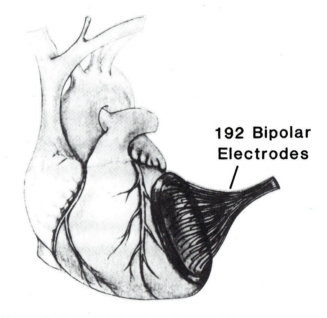

192 Bipolar Electrodes

Figure 5.6. Top: Photograph of two "egg" electrodes, each bearing 480 contacts 3 mm apart, for endocardial mapping in the dog atrium. From Allessie et al. 1982. Bottom: Placque electrode array used to map activation on the epicardial surface of the dog ventricle, bearing 384 unipolar electrodes at intervals of 2 mm. Adapted from Wit et al. 1982 by permission of the authors and the *American Journal of Cardiology*.

tionary array of 60 electrodes, physiologists M. J. Janse and Frans J. van Capelle, working with ischemic regions of pig heart in the University of Amsterdam's teaching hospital, obtained the isochronal maps of ventricular fibrillation at spatial resolution of about 1.5 mm [Janse et al. 1980]. They show diverse patterns, sometimes resembling many little wavelets irregularly pirouetting about 5-mm paths (Figure 5.8), as though the early stages of fibrillation were merely an aggravated case of two-dimensional reentry: a dense clustering of interacting rotors.

Distinguishing Vortex Reentry from Ring Reentry

This two-dimensional mode of reentry (the vortex-like "rotor") is distinguishable conceptually from one-dimensional circulation of a pulse on a loop. It may also be distinguishable in the laboratory by its different response to drugs that alter the rate of rotation [Allessie et al. 1977]. A. K. Grenader and colleagues [Grenader and Zurabishvily 1984; Agladze et al. 1983] contrived a clinical test. Their test, like Allessie's, depends on the qualitatively different mechanics of the two kinds of reentry. The cycle period in each case depends on such membrane properties as the duration of the refractory period, the conduction speed, etc., but in different ways. Those factors, in turn, depend on the concentration in the patient's blood of cardiac pharmaceuticals such as lidocaine. (Lidocaine plugs the fast sodium channels and thus diminishes propagation speed; it does so selectively in depressed cells, tending to remove them completely.) By varying the dose and watching how the dysrhythmia's period varies in response, a doctor should be able to discern which kind of reentry underlies it. Quite different treatments are required, depending on that diagnosis.

Rotors and Phase Singularities

How is any of this connected with phase resetting or phase singularities? First of all, it needn't be. There are uncountably diverse mechanisms by which action potentials, especially slow-response action potentials, in heart muscle may become reentrant. But for the sake of following our theme of phase singularities from circadian rhythms to neural and cardiac pacemakers through rotors in heart muscle to the three-dimensional equivalent in chemically excitable media, I describe one potential mechanism of rotor creation that is intimately associated with phase resetting.

The idealized nonwandering rotor has a pivot point. The immediate neighborhood of the pivot is arrhythmic, but fibers one space-constant away are rhythmic with the common period of rotation. Isochronal contours (successive wavefronts) converge near the pivot. The pivot, in other

Box 5.C: The State of the Art

Nerve networks share the essential properties that make heart muscle vulnerable to rotating waves. In one particular nerve network—in the colonial coral polyp *Renilla* (the "sea pansy")—those waves are plainly visible and they are slow enough to photograph (only about 5 cm/sec in contrast to the 50 cm/sec of ventricular myocardium). Waves passing along the coral's surface stimulate a green bioluminescent flash as they go, revealing both "ectopic foci" and "reentrant arrhythmias" [Buck 1973]. In other excitable epithelia, action potentials also travel at convenient speeds in the order of 10 cm/sec through a uniform isotropic medium that would readily lend itself to experiments with two-dimensional reentry [Josephson and Schwab 1979; Anderson 1980; Schwab and Josephson 1982]. But heart muscle presents a far less ideal opportunity to the experimentalist.

Electrical impulses travel at speeds up to half a meter per second in heart muscle, invisible but for the muscular contraction that trails behind. In tachycardia the muscle never gets a chance to relax fully in the wake before the next activating wave is upon it. How then is one to discern the spatial pattern of activation, the normal or abnormal arrangement of isochrons? One of the first major efforts entailed reviving a human heart shortly after its owner perished of other causes. Impaled with 870 electrodes about 7 mm apart, the beating heart provided electrical information from which to reconstruct three-dimensionally the normal sequence of activation [Durrer et al. 1970; this was later done also with the somewhat different hearts of human infants: Brusca and Rosettani 1973]. But what happens when the heart must not be further damaged or when the rhythm is tenfold quicker than usual, or very fine-grained?

This problem must be solved; nothing tells a cardiologist what is going wrong in a sick heart more clearly than an isochron map. Accordingly much effort is currently invested in automation of more and more detailed mapping techniques for use in the operating room and in animal experiments [Janse et al. 1980; Boineau et al. 1980a; Harumi et al. 1980; El-Sherif et al. 1982; Janse and van Capelle 1982; Smith et al. 1982; Wit et al. 1982; de Bakker et al. 1983; Mehra et al. 1983; Witkowski and Corr 1984; Allessie et al. 1984; Cardinal et al. 1984; Downar et al. 1984; El-Sherif 1985; El-Sherif et al. 1985; Allessie et al. 1985]. For the present, at most 700–1000 electrodes are deployed, with spatial resolution limited to about 2 mm (almost adequate in normal tissue with space constant of ½ to 1 mm, but still inadequate in the ischemic tissue that commonly initiates arrhythmias). Several groups are beginning to systematically deploy arrays of plunge needles bearing closely spaced electrodes to reveal the three-dimensional anatomy of both transient and sustained arrhythmias.

Another approach toward the goal of recording the pattern of fast electrical activity in a living heart begins with an entirely different technology. Instead of directly picking up the signals by touching the heart with a metal pinpoint, would it be possible simply to look, essentially photographing the voltages? Not quite, but almost. Cell membranes painted with a voltage-sensitive dye will fluoresce when illuminated by strong light. The fluorescence depends on the local voltage. Sawanobori et al. 1984 and Hirano et al. 1982 monitored reentrant tachycardia in a ring of frog atrium in such a way. For better resolution one might scan the heart with a pinpoint of blue laser light and record the fluctuating fluorescence. Meanwhile, just as in a scanning electron microscope, scan the face of a TV screen at the same rate, lighting up each point in proportion to that fluorescence. If you scan fast enough, you paint an essentially instantaneous picture of voltages on the surface of the heart, and the picture changes as the voltages change. This technique has been brought to the point of routine high-resolution reliability by Steven Dillon and Martin Morad in the Department of Physiology of the University of Pennsylvania Medical School: every 12 milliseconds they can record the voltage at 4096 points on the exposed muscle surface [Dillon and Morad 1981]. Very detailed maps of complicated activity erupting onto the epicardial surface are about to become available.

Moreover, similar arrangements may also provide the most flexible means of patterned stimulation for deliberate instigation of arrhythmias. By infusing the preparation with a dye that adheres to the membrane, one makes cells photosensitive [Farber and Grinvald 1982]. If an appropriate image is then projected on the epicardium, it may be feasible to create controlled gradients of depolarization or hyperpolarization along the surface [Scott Fraser, personal communication]. This is not so easily done with electrode arrays because extracellular current penetrates membranes in both directions in different parts of each muscle fiber.

However the only foreseeable technique for observing the three-dimensional structure of an arrhythmia remains the array of plunge needles [Witkowski and Corr 1984].

words, has become a phase singularity much as in the phase-resetting protocol that we called a "pinwheel experiment." Much as, but not exactly as. Two crucial differences are conspicuous: In the foregoing examples cells are indeed rhythmically active but not *autonomously* rhythmic: their rhythmicity derives from the fact that they are mutually coupled. In the pinwheel experiments of prior chapters, cells were autonomously rhythmic and independent.

For mathematicians this contrast makes all the difference in the world. Empirical scientists, however, sometimes take leaps of faith from one

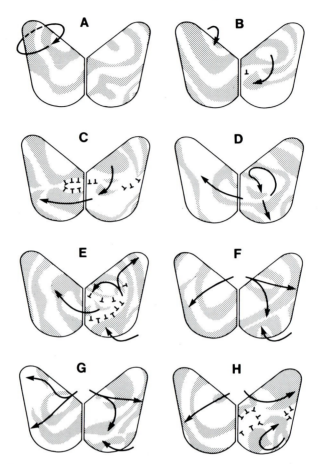

Figure 5.7: Consecutive isochronal maps (A through H: covering 600 msec by sixty 10-msec intervals, alternately white and stipple), from activations detected by the electrode array of Figure 5.6 (top). Activity in the upper surfaces of the two atria is shown during transition from regular rotation to fibrillation. The small T's indicate loci of failed propagation (perhaps still refractory from prior activity). Adapted from Allessie et al. 1982. The original publication from which this figure was adapted uses a discrete color code, not to be confused with the use of continuously graded color throughout this book to represent phase in a cycle.

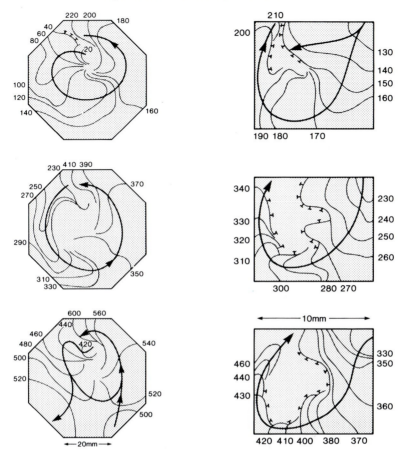

Figure 5.8. Left, hexagons: Three successive turns of one meandering activation front in pig left ventricular epicardium (previously ischemic) during a 5-Hz tachycardia. Activation sequences were derived from 60 microelectrodes spaced 4 mm apart within the area shown. From Janse et al. 1980. Small T's indicate failure of further propagation. Times are in milliseconds from arbitrary beginning.

Right, rectangles: Tighter, quicker, and less complete rotation of three successive activation wavefronts is indicated by isochrons sketched on a 1-cm² area of fibrillating pig ventricular epicardium (previously ischemic). Activation sequences were derived from 60 microelectrodes spaced 1.5 mm apart within the area shown. Wave speed is about 0.4 m/sec in both cases. From Janse et al. 1980 by permission of the authors and the American Heart Association, Inc.

phenomenon to a similar one, carrying only a fine thread by which they hope to haul heavier cables across later to build a proper bridge. Empirical scientists sometimes fall into a bottomless void. The objective of the following chapters is to show that there is in fact solid ground on the far side of this chasm and that it is reachable by an unsupported leap.

CHAPTER 6

Time Encircles a Singularity

And here, "... my dear Watson, we come into those
realms of conjecture where the most logical minds
may be at fault: each may form its own hypothesis
upon the present evidence, and yours is as likely to
be correct as mine."

—Sherlock Holmes, in
"The Adventure of the Empty House"

The intuition of physiologists Garrey and Mines before World War I stands abundantly confirmed: arrhythmias that typically precede fibrillation can be (and commonly are) triggered by an ephemeral stimulus (believed to come somehow from the central nervous system), and the aftermath somehow involves rotating waves. But we have not yet answered the question, "What is the connection between singular stimuli and rotating wavelets?"

To find an answer, we must first recognize that the mathematics of phase singularities is both more general and less revealing than we may have imagined in the first flush of excitement over finding it. It is more general because the arguments do not depend on spatial uniformity of timing: the topology works the same for a spatially distributed complex oscillator, like the sinoatrial node, as it does for a single cell. But the mathematics is also less revealing because the ambiguity of timing implied by the coloring theorem is no more than that: the theorem only tells how to find a singular stimulus in whose wake the oscillator cannot spontaneously and predictably return to its usual mode of beating. We cannot learn from topology alone what kind of arrhythmia is conjured up by the singular stimulus, but only that there must be a physiologically reasonable stimulus (larger than those that elicit odd resetting) after which one cannot count on reliably timed return to the usual rhythm. To find out how and why timing is unreliable after that stimulus, we must bring together the clue involving time (Chapter 3) and the clue involving space (Chapter 5).

The essence of a beating heart, after its rhythmicity, is the spatial arrangement of that rhythmicity: the timing relations among parts. Chapter 5 gave several strong hints that fibrillation grows out of a distinctive spatial arrangement of timing. Let us then guess that spatial nonuniformity is the missing ingredient: that some arrhythmias, possibly including the initial stages of fibrillation in real hearts, do indeed develop from vortex-like waves triggered somehow by a rare singular stimulus striking in that critical phase. We must think about timing in its physical context: we must think of space and time together. That is not so hard to do if spontaneously oscillatory tissue is retained as our model for now, and our reservations and compunctions about the fact that ventricular muscle is not spontaneous in its repetitive firing are swept under the rug (see Appendix).

Phase Resetting in a Neighborhood: Putting Space and Time Together

The missing link in the argument from phase singularities to rotating waves lies quite simply in the color wheel encircling each singularity in Figures 4.2, 4.4, and 4.9. These colors can be traced back along isochromes to the horizontal axis, where the stimulus is too small to reschedule the timing of heartbeats noticeably. There we know that, by definition, latency decreases through one full cycle as the coupling interval increases through one full cycle. Along any closed path around a phase singularity, then, latency also runs full cycle.

What does this mean for a real stimulus, e.g. electrical shock, a volley of nerve impulses arriving simultaneously over a wide area from outside the heart, or a temporary local loss of blood supply? It turns out to mean that the singular stimulus sets up the perfect pattern of latencies to start a rotating wave where there was none before. To see why, we must play a game with colored pencils. It begins again from the parochial perspective of the single fiber.

Consider the sinoatrial node as a hypothetical example. The human node is a region only 15 mm × 5 mm in area and 2 mm in thickness. But functionally its size is comparable to the rest of the atrium or the ventricles because, lacking fast sodium channels, nodal cells give only a "slow response" and conduct excitation an order of magnitude slower than other parts of the heart. The node is composed of hundreds of thousands of individually rhythmic pacemaker cells, innervated diffusely by the vagus nerve. (They are commonly called "cells" rather than "fibers" because they are much less elongated than the working myocardial cells in the more muscular parts of the heart.) Some parts of the sinoatrial node normally fire as much as 50 msec in advance of others: there are gradients of timing that change according to circumstances [Boineau et al. 1978; Boineau et al. 1980b; Bleeker 1982]. There are also gradients of stimulus intensity during vagal activity, due to regional variations in the abundance of vagal nerve endings. At the moment when a stimulus arrives, the many cells of the node are found at slightly different stages of their cycles. The same is true on a larger scale within the atrial muscle outside the sinoatrial node, because it is activated progressively away from the node by a moving wave and is diffusely innervated from many sources. On a still grosser scale, the same may even be said of the whole heart.

Cells in different places will feel the stimulus at different strengths: no stimulus strikes its whole area of impact uniformly. Let us suppose that in some little region of heart muscle such as the sinoatrial node, stimulation varies from strengths below a phase singularity to strengths above it, within a space of a centimeter or so. Within some central area the stimulus is strong enough to elicit even resetting. This region is necessarily bounded, like any island, by a ring-shaped beach, just beyond which the weaker stimulus elicits only odd resetting.

Figure 6.1 gives some idea of the typical situation. Here we have a chunk of tissue with two successive waves sweeping across it in an orderly way. This much is deliberately exaggerated: usually[1] there is at most one excitation wavefront on the heart, and within any smaller area only a fraction of a cycle is represented at any one moment. But to stress the essential feature that there is usually some gradient of timing, it is convenient to sketch a gradient full-cycle, from one wavefront to the next. Within this area cells at equal distances behind the wavefront are at nearly the same stage of the cycle: past positions of the wavefront are *pre*-stimulus isochrons that will be restructured by local phase-resetting in response to a stimulus, as we are about to see by plotting *post*-stimulus isochrons in Figure 6.2.

Suppose a depolarizing stimulus occurs with epicenter not too far from the region that is presently at the singular phase for depolarizing stimuli. This is the critical isochron, a past position of the activation front. The impulse might be a broadside from the vagus nerve ramifying in the atrium and so causing local phase resets; it might be a wavefront encroaching out of the depths of the muscle; or it might be current from an extracellular electrode dispersing through a wide region.

---

[1] During the first seconds of fibrillation such steep gradients may actually occur in ventricular myocardium; using an array of 28 electrodes on dog epicardium Ideker et al. 1981 show one such case involving a 9-Hz ventricular tachycardia. See also Ideker et al. 1984 and Bardy et al. 1983.

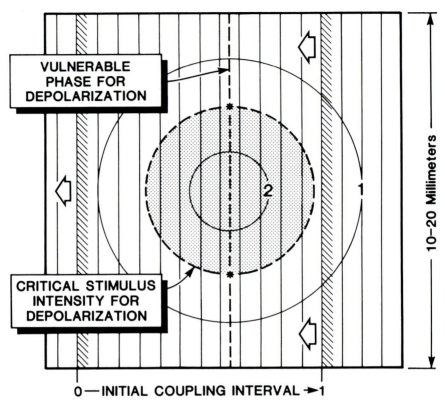

Figure 6.1a: To make the central message plainer, this diagram exaggerates the orderliness both of nonuniform stimulation and of the pre-stimulus dispersion (gradient) of phases present within a small area. For the sake of displaying the whole range of phase, two wavefronts are shown (shaded behind), their past positions marking the pre-stimulus isochrons in finer black. The critical isochron is shown in heavy dashes near the mid-cycle completion of repolarization (where the depolarizing singularity or discontinuity is commonly found in experimental preparations (Chapter 4). A stippled island of strong stimulation, also centered mid-cycle for convenience here, is surrounded by a heavy dashed border that separates regions of even and odd resetting. Cells at the intersection of heavy dashed curves receive the singular stimulus.

In Figure 6.1 we also see concentric rings surrounding the shaded strong-stimulus island. The contour rings represent beaches of decreasing impact further from the center. Rings 2 and 1 might represent current strengths above and below the depolarizing singularity on Figure 4.9.[2] The heavy dashed ring between them is the beach separating the strong (even-resetting) center from the weak (odd-resetting) periphery. The area

[2] Stimuli provided to a tissue of cells by external current are harder to quantify than those delivered to a single cell through an intracellular electrode or to a tissue by neurotransmitters. The net current though a closed membrane is, of course, zero, since it must both enter and exit. But the net *effect* is apparently non-zero and on the average depolarizing,

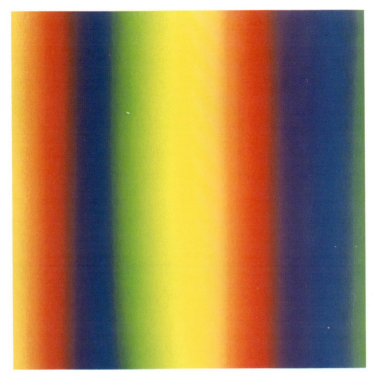

6.1b: The pre-stimulus isochrons of (a) are color-coded, red representing action potential and purple-blue the refractory period.

of muscle between rings 1 and 2 then spans all coupling intervals and the critical range of stimulus strengths in an orderly way. Within this area all combinations of stimulus strength and coupling interval near the phase singularity are laid out on the heart much as they are in Figure 4.9.[3] Thus the post-stimulus isochromes of Figure 4.9 can also be painted on this patch of muscle, along with their zone of convergence: the singularity appears where the heavy dashed critical isochron (the pre-stim-

---

and more so if the current is greater, regardless of its polarity. Stimuli that are hyperpolarizing in net effect apparently cannot be delivered via uniform extracellular electric fields.

[3] Taking literally the two full cycles of phase here depicted, this diagram suggests propagation speeds as low as 0.01 m/sec, commonly obtained only in the sinoatrial node, atrioventricular node, or transversely to fiber orientation, or in areas of ischemic depression and conduction block. Propagation elsewhere is fifty to a hundred times this fast. To apply this diagram to such tissues, remember that we are really talking about the hundredth of a full cycle when the critical range of coupling intervals is represented within our centimeter-sized patch.

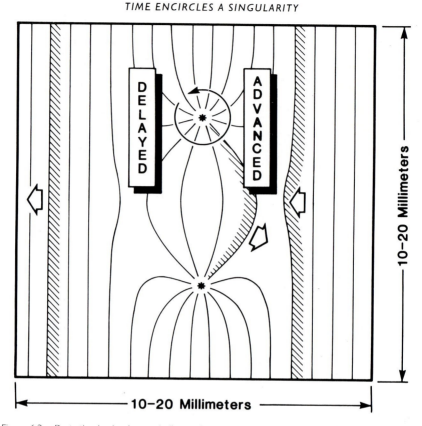

Figure 6.2a: Post-stimulus isochrons similar to those in Figure 4.9 are computed from the simplest qualitatively correct model (see Appendix) and mapped onto Figure 6.1 with pre-stimulus phase horizontal and stimulus size falling off radially from center with 0.5–1.0 mm space-constant. The immediate-firing isochron (0) arcs from one singular point to the other. The two endpoints of this new segment of wavefront lie in vortices of latency, necessarily reduplicated in mirror image because stimulus intensity falls off from its central maximum in both directions along the vulnerable phase isochron.

ulus isochron of the singular phase) crosses the heavy dashed critical-impact beach (Figure 6.2).

Here, once again, we have the "pinwheel experiment," but now twinned. Before the arrival of an outside stimulus, isochromes were smooth contours without endpoints encircling the sinoatrial node, the source of normal activation in each heartbeat (Figure 5.4). When the stimulus arrives, each intensity level (in particular the beach bounding an island of strong stimulation) is a closed ring (closed unless the island overlaps an inert region such as the atrioventricular boundary). Closed rings, however complicated their shapes may get, will intersect a wave-

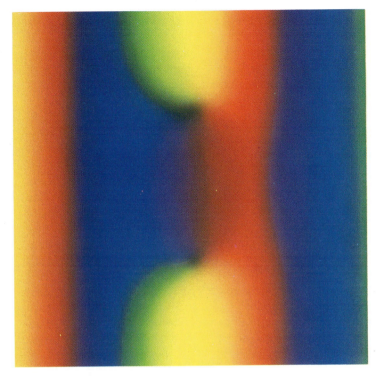

6.2b: The same presented as color contours without other busywork.

front at an even number of places: not at all or at least twice. So unless the stimulus is centered near a boundary, we expect to see any part of the rainbow latency diagram painted twice on the heart if it is painted at all: in particular, the phase singularity and its multihued environs should commonly appear in (distorted) mirror-image pairs. (For a reinterpretation of Figures 6.1 and 6.2 in terms of refractoriness, conduction block, and excitation thresholds, see Figures 7.13 and 7.14.)

Rainbow Neighborhoods Rotate

What do these colors mean, when visualized as though actually painted on the muscle itself? They mean that paired vortices of electrical activity have been started in this sheet of independent and independently reset pacemakers! Each isochron color runs through fibers that will fire simultaneously: the ones with the same latency after the stimulus. In particular the zero isochron is the locus of immediate firing: the new wavefront. Near each phase singularity these latencies decrease in sequence

around a circle: so the stimulus has created a rotating sequence of firings—in fact, a mirror-image pair, one clockwise and one anti-clockwise,
both turning at exactly the period of spontaneous rhythmic firing in these
isolated pacemakers. It is hard to imagine a better set-up for starting a
reentrant wave. But what determines the wave's initial speed? It must
have something to do with the spatial gradients of reset phase.

Behavioral Discontinuities

When the equations of electrophysiology were first deployed to illuminate
phase resetting, continuity of even-resetting curves was far from evident.
Fault lines in the contour map are common, even the rule [Best 1979;
Chay and Lee 1984; Glass and Winfree 1984; Michaels et al. 1984; Chay
and Lee 1985; Reiner and Antzelevitch 1985]. In order to verify strict
mathematical continuity, computer experiments had to be carried to more
decimal places of precision than one could approach in the lab. In other
cases the continuous equations of electrophysiology in fact describe
strictly discontinuous phase-resetting curves. Still more unexpectedly, the
near-discontinuities and discontinuities were commonly not near the sudden upstroke of the action potential, where one might plausibly have
expected abrupt change. Correspondingly, in the laboratory physiologists
do commonly find big vertical jumps in the phase-resetting curves of
typical pacemaker cells and, as in the scouting computations, often not
near action potential time [see Box 3.A and many references in Winfree
1980; Glass and Winfree 1984; Guevara 1984].

What that means in the context of spatially distributed resetting is that
nearby places may have very different phases. If adjacent fibers fire independently without triggering their neighbors, they must be expected
to fire spontaneously at considerably different times: a wave of simultaneous firing would appear to creep across the earthquake fault line of
local phase, even standing still before leaping forward again from the far
side of the crack at a speed proportional to the flatness of the local phase
gradient.

This unrealistic inference[4] depends on imagining—as we have done up
to now, just to make the bookkeeping easier—that every cell's timing is
independent of its neighbors'. In reality uniform propagation prevails due

---

[4] How unrealistic is it? Islands of "functional conduction block" somehow arise and
careen wildly across the landscape of the heart during arrhythmias. These are virtual
discontinuities in the timing of activation, where phase gradients become extremely steep.
They are places where time breaks down, very similar to phase singularities except that
they are often bigger. There is at present no good understanding of their origin and behavior.
A line of conduction block often intrudes, *deus ex machina*–fashion, to terminate reentrant
arrhythmias in isochronal mapping experiments. Allessie et al. 1985 show a flutter wave
circulating around a 5-cm long arc of conduction block in the healthy atrium of a dog,

to an overlooked physical principle: electrotonic coupling. Up to now we have seldom dealt explicitly with the coupling between adjacent cells. It cannot be ignored any longer. The coherent rhythm of the sinoatrial node, for example, is due to mutual synchronization among its many cells through electrotonic currents [Jongsma and van Rijn 1972; Sano et al. 1978; Jongsma et al. 1983]. Propagation of waves across the atrial muscle, for another example, is due to electrical conduction; without electrical coupling the heart could not maintain its coordination. The electrical junctions between consecutive fibers and, to a lesser degree, the spread of current between adjacent fibers make it difficult for contiguous patches of healthy membrane to remain long in dissimilar electrical states. Similar fibers sufficiently close together (1–2 mm) will thus do about the same thing at about the same time, smoothing over the erstwhile discontinuity [Fel'd and Morozova 1979; Medvinsky and Pertsov 1979; Joyner et al. 1983].

For topological reasons, the phase singularity inside a circular gradient of phase is immune to such resolution. A circular gradient cannot be smoothed over. The only way it can vanish is by fusion with a mirror-image gradient or by migrating to some boundary. Once set up in a sheet of pacemaker cells, a circular pattern of timing seems likely to persist. The way in which it persists, however, depends on the electrical coupling between neighboring fibers [Jalife 1984; Jalife and Michaels 1985]. If adjacent cells were not electrically coupled at all, this circular timing gradient would in principle persist, cycle after cycle, until distorted beyond recognition by the cumulative effect of small differences in natural period between independent cells. But given some mutual coupling, neighboring cells trigger one another, maintaining local coherence. Given such coupling it may not even be necessary for cells to fire spontaneously (which, in fact, only nodal pacemaker cells and Purkinje fibers do in a healthy heart). Simulations reveal that in case of reentry, neighbor coupling (diffusion of electric potential, analogous to diffusion of chemical concentrations, to be considered in the next chapter) in an excitable medium—whether or not spontaneously periodic—serves much the same purpose as spontaneity: it keeps the wave rotating [Winfree 1974b; Winfree 1978; Winfree 1980; van Capelle and Durrer 1980]. The same is observed in a chemical excitable medium (Chapters 7 and 8): neighbor

---

much as in the forecasts of Balakhovsky 1965; but why it has that length or shape, no one can say. El-Sherif et al. 1982 demonstrate others 3 to 11 cm long in dog ventricle, diversely shaped, sometimes with the middle segment vanishing to leave only the ends as foci of counter-rotating vortices. As a first step toward understanding such loci, this book attempts to interpret only situations in which they are as small and restricted as possible: mere points, not lines, of phase singularity. Without attempting more, we will see in Part III that the three-dimensional arrangement of these punctate singularities is subject to definite laws.

coupling stabilizes rotation around phase singularities even in areas that would never fire without provocation. Still, these are only metaphors. What about real heart muscle?

Cardiac Rotors Created by a Single Stimulus

The experiments cited in Chapter 5 showed that rotors persist in heart muscle like hurricanes on the Caribbean sea, independent of any hole or ring to circulate around, as though owing their geographic location only to the location and timing of some special stimulus. The earliest demonstration by Allessie and colleagues [Allessie et al. 1973; Allessie et al. 1976; Allessie et al. 1977] in fact used a premature stimulus to create the reentrant focus. This rotor turned at a period longer than the minimum refractory period but still short enough to seize control from the normal pacemaker: 7 to 10 cycles per second, compared to the rabbit sinus node's 1 to 3 cycles per second. Each rotor may be located anywhere, but wherever it is, it occupies about a square centimeter. Smaller (unstable) rotors appear even within the sinoatrial pacemaker node itself, turning within the space of about a square millimeter [Allessie and Bonke 1978]. But the rotors found first were not confined to the region of spontaneous automaticity; they seemed perfectly stable in the surrounding atrial muscle, which is excitable but not spontaneous.

This is essentially the fulfillment of what Russian biophysicists had attempted to demonstrate in frog atrium [Rozenshtraukh et al. 1970], and had computed in anticipation of more adequate laboratory techniques [Gul'ko and Petrov 1972; Shcherbunov et al. 1973]. In dog atrium, fibrillation induced by a single well-timed stimulus apparently consists of several reentrant wavelets coexisting and interfering with one another enough to spoil any semblance of rigid rotation [Allessie et al. 1985].

In the dog ventricle, J.M.T. de Bakker and colleagues [de Bakker et al. 1979] were able to demonstrate the activation sequence during the initial stages of fibrillation induced by a single stimulus in the ventricular vulnerable phase. A wave rotated about a moving pivot for two complete cycles before moving outside the 30-mm grid of electrodes.

It should be noted that rotors in heart muscle are seldom very symmetric. The rotor's core is often so elongated that it is described as "an arc of functional conduction block" [Mehra et al. 1983; Allessie et al. 1984; El-Sherif 1985]. This[5] asymmetry may derive largely from the asymmetry of neural excitability, even in a perfectly isotropic material: see the Appendix and see also how elongated the singularity appears in Figures 4.1 to 4.5 and 4.9. But it must also be influenced by the inherent asymmetry of a fibrous material like cardiac muscle, conducting 3–4

[5] See note 4.

times more rapidly along the grain than across it [Dillon et al. 1985; Spach and Dolber 1985].

Symmetric or not, this is not a good situation. Reentry typically repeats at very much shorter intervals than the spontaneous firing of the sinoatrial node, so that tissue harboring a rotor pair is a source of tachycardia. Worse, the prerequisites for some new random stimulus to create a new rotor are now much less demanding, with every phase of the cycle simultaneously always present. The timing of a stimulus is then no longer critical: whenever it arrives, it will somewhere find a strip of tissue (an isochrome) in the critical phase. Wherever along that strip the stimulus strength rises or falls through the critical range, there yet another rotor is born. This is the dilemma discussed by Chen et al. [1986a] after their epicardial mapping of attempted defibrillation: the would-be defibrillating pulse finds some tissue in the vulnerable phase and so must be very strong or very weak lest it reinitiate fibrillation.

And still worse: even the normal human heart is laced with organic discontinuities [Spach et al. 1981; Spach et al. 1982; Spach and Kootsey 1983; Spach and Dolber 1985]. They seem capable of shattering rotors, and so shattering isochromes into many little wavelets [Krinsky 1978]. This may be how the rotor implicit in a phase singularity ushers in more and more reentrant wavelets, culminating in the turbulent lethal arrhythmia called fibrillation.

How might a single stimulus inaugurate this avalanche of disasters? There are many ways; here only the one class of candidates that appear to involve phase singularities will be discussed. To preview briefly, it turns out that the latency diagram for a tissue can be essentially the same as for a single fiber if the tissue is very small and uniform or stimuli are remote from the singularity (very strong, very weak, or given during the refractory period). But if the tissue is large and nonuniform then the latency diagram of the extended tissue changes near the singularity: the singular point enlarges to become a black hole. The black hole contains all stimuli that create a pair of singularities somewhere within the tissue. The prospect of arrhythmia now gapes open as a substantial target for otherwise innocuous but untimely stimuli, whose arrival during this window of vulnerability implants seeds of incoherence in the tissue. We thus arrive at the surprising conclusion that a complicated sizable biological clock, even if composed of cells whose singularities are innocuous hueless points, cannot be made without an Achilles' heel: a substantial black hole of vulnerability.

Black Holes of the Heartbeat

Let us now think again about a gross stimulus affecting a substantial region of periodically active tissue susceptible to implantation of a phase

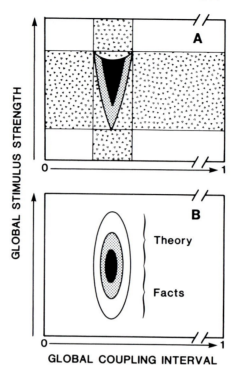

Figure 6.3: (a) On rectangular coordinates similar to all prior latency diagrams, we plot the onset time and some measure of the strength of a stimulus administered to a finite continuum of oscillators, each of which feels it at a slightly different local time and strength. Within the coarsely stippled range of times, some local oscillators will feel the stimulus at critical phase. Within the coarsely stippled range of strengths, local strength will be critical for some oscillators. Within the area of overlap (exaggerated to make room), grey and black shading denotes stimuli that create singular conditions somewhere within the tissue. Within the finely stippled fringe, singularities may be paired too close together for stable reentry. (b) An electrical stimulus applied to the normally beating heart will induce fibrillation if it falls within a narrow window of time called the "vulnerable phase" and if it is strong enough but not too strong. As those criteria are approached, premonitory extra beats are observed: one extra beat following any stimulus above the bottom U-shaped contour, or several following a stimulus in the stippled region above the next higher U. Fibrillation follows a stimulus in the black region above the innermost U. Closure of all three contours around the top, completing a "target with bull's-eye" is conjectural, based on the supposition that a big enough stimulus, even during the vulnerable phase, would leave the entire mass of muscle uniformly depolarized. Adapted from Winfree 1983. (Shibata et al. 1985ab, 1986, and Ideker and Shibata 1986 seem to have confirmed this. Fall 1986 addendum: Dr. Ideker has since discovered that this was previously established experimentally; see A. Fabiato, Ph. Coumel, R. Gourgon, R. Saumont [1967] *Arch. Mal. Coeur* 60 [4]:527-544; and C. Lesigne, B. Levy, R. Saumont, P. Birkui, A. Bardou, B. Rubin [1976] *Med. Biol. Eng.* Nov.:617-622.)

singularity. Let the stimulus be applied at a moment measured from the peak of the electocardiogram's R wave: call that the global coupling interval to distinguish it from the local coupling interval, at which a given cell feels the stimulus. And call the intensity of the applied stimulus the global stimulus strength to distinguish from the local stimulus strength felt by each affected cell according to its anatomical position.

Rule a piece of graph paper with global coupling interval from 0 to 1 horizontally, and vertically with global stimulus strength from 0 to something big enough to essentially resynchronize the whole of a given piece of tissue (Figure 6.3a). The coordinates resemble those often used before, e.g. in Figures 4.1 to 4.5 and 4.9, but now we are not going to color each point according to post-stimulus latency. Instead we will blacken any point (stimulus) that initiates an arrhythmia by the criterion that it creates a phase singularity somewhere in the tissue. Anticipating the outcome, we will discover a solid black disk, a "black hole" in this diagram.

The rules are as in Figures 6.1 and 6.2:

(1) The tissue initially contains a phase gradient. We take special interest in the critical pre-stimulus isochron, if it appears within the tissue at the moment of stimulation.

(2) The tissue receives the stimulus in a graded way. We take special interest in the beach bounding a central island of even resetting, if the stimulus is strong enough to produce one and not that strong throughout the whole tissue.

(3) We ask what stimuli result in those two special loci intersecting. Points of intersection (phase singularities) are assumed to develop into rotating wave sources if not too close together.

In seeking intersections [rule (3)], rule (1) tells us to forget about coupling intervals outside a vertical band before which the critical pre-stimulus isochron has not yet appeared in the tissue and after which it has already run through (Figure 6.3a). During normal sinus rhythm in the human heart this band could be as wide as 80 milliseconds, the total interval from firing of the atrioventricular node to activation of the last fiber of ventricular muscle. It may be even wider, to the extent that fibers, once triggered, complete their cycles (repolarize) at diverse rates. And during any reentrant arrhythmia, with every stage of excitation/relaxation always present, this band occupies the whole cycle.

Rule (2) tells us to forget about stimuli that are too small or too big, leaving only the intermediate horizontal band of Figure 6.3a. The much larger stimuli are in fact commonly used to eliminate reentry during cardioversion (see Box 6.A).

Box 6.A: The Defibrillator Backfires

Patients hospitalized for cardiac intensive care often fibrillate sponta-neously. In fact, doctors often provoke fibrillation deliberately in order to find out what kind of stimulus is causing the trouble and what medication best prevents it. And fibrillation is a routine step in the recovery of normal rhythmicity during the last stage of open-heart surgery. In a heart as big as Man's, this electrical turbulence is perniciously persistent. It can be fatal within a few minutes. "Defibrillating" equipment is kept at hand to restore synchrony by a massive electrical countershock. Needless to say, cardiol-ogists like to use a 5-kilovolt machine as gently as possible to avoid tissue damage such as one reads about in the first accounts of electrical de-fibrillation [Abildgaard 1775]:

> I tried the electric shock directed through the breast to the dorsal spine ... [the hen] raised up suddenly and quietly walked on its feet.... I tried this same experiment with a poultry-cock which, after many [Leyden jar] shocks had been directed to its head, appeared altogether dead so that blood ran forth from the nose and mouth, but from the shock on the breast it quickly flew away and knocked the electric vessel on the earth and broke it.

Gentler use is called for, for example, when the objective is merely to erase a benign dysrhythmia like atrial flutter. The defibrillating thunderbolt-machine was accordingly scaled down and called a "cardioverter" by its physician-engineers in Bernard Lown's laboratories at Harvard. But a counterintuitive danger was discovered in this gentler application, a danger seldom experienced in full-strength applications: impulses of the right size, inadvertently administered during the vulnerable phase, sometimes trig-gered an episode of fibrillation!

The "right size" is practically important for defibrillation. If fibrillation consists of action potentials propagating in normal tissue but in abnormal spatial and temporal patterns, then Figure 6.3 may be pertinent for esti-mating that right size. During rapid ventricular tachycardia, at every mo-ment some tissue is always in the vulnerable phase of its local excitation and recovery. If the local stimulus size is in the black region, then fibrillation may be instigated in that tissue even while tissue elsewhere is being de-fibrillated. If the black region had no top, it would be impossible to de-fibrillate by a single shock. Thus it should be no surprise that there is a top. Successful defibrillation, then, may require either of two arrange-ments:

(1) that even the most heavily electrified tissue must lie below the black hole's bottom, while even the least electrified regions are all depolarized;

the danger in this case is that some regions may escape stimulation from a single shock; or

(2) that even the least electrified tissue must lie above the top of the black hole. The danger in this case is tissue damage, particularly if the applied shock is not uniform and so electrifies some regions very much more than others.

Intermediate stratagems may initiate fibrillation in one place even while they terminate it in other places. Chen et al. [1986ab] interpret failures of defibrillation in such terms. Recent studies by Shibata et al. [1985ab, 1986] and Ideker and Shibata [1986] measure the black hole and its target-like fringes of extrasystoles.

Consider now the rectangular box where the horizontal and vertical bands overlap. It contains stimuli that place a beach of critical intensity in the tissue while the critical isochron is racing across it.

Intersections can be created anytime during the interval while the critical pre-stimulus isochron (following in the wake of the phase 0 isochron, the action-potential wavefront) is traversing the presumptive ring of critical stimulus intensity (Figure 6.4). That interval, a measure of pre-stimulus nonuniformity of phase within the stimulated tissue, is also proportional to stimulus-ring diameter: zero for stimuli at the bottom of the little box and greater for stronger stimuli. Thus the bottom edge of the black region is V-shaped in Figure 6.3a. At maximum stimulus size there might remain only a few isolated corners of the tissue (such as the atrial appendages or parts of the ventricle bounded by the atrial border or by an infarct) within which intersections can occur. Intersections would occur in isolated regions at discretely separate times (Figure 6.5). In such situations, the top edge of the black region might be convex as in Figure 6.3b or might even be concave as in Figure 6.3a, with horns tapering to nothing. But the main feature emerging from this argument is that there will be a narrow zone of vulnerability with a V-like bottom and a finite top.

The fringes of the black hole represent glancing intersections, nearly tangential, with the two curves crossing then uncrossing again nearby. Rule (3) suggests that such paired mirror-image singularities may be ephemeral, the oppositely reentrant waves recombining with alacrity. Therefore erase the black within a shaded fringe of some depth, where only a few ectopic beats might be expected; lasting arrhythmia occurs only within the remnants of solidly blackened interior. The V-shaped

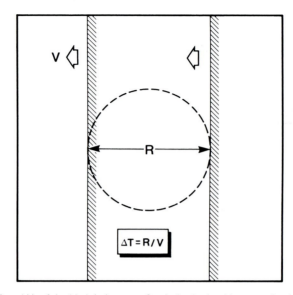

Figure 6.4: The width of the black hole at any fixed stimulus level is proportional to the range of phases R inside the affected tissue (the island of even resetting). In the simplest case, a smooth phase gradient and a circular island, the interval of possible intersections, ΔT, is the island diameter divided by the wave speed V.

bottom may thus soften to a U shape within the fringe of unstable reentry. This fringe may completely envelop the black hole as in Figure 6.3b [Winfree 1983], but need not appear on concave borders, e.g. atop Figure 6.3a. The fringe may also penetrate completely to the center of the vulnerable area, leaving no core of blackness, in masses of muscle so small that counter-rotors cannot be far enough apart to persist (e.g. rabbit hearts, incapable of sustained fibrillation).

Figure 6.6 shows typical measurements of the vulnerable region on such a diagram, using electrical stimuli on dog and turtle ventricle [from Brooks et al. 1955]. The detailed appearance of the region varies from one technical publication to the next, likely due to different electrode arrangements. But there is no doubt of a V-shaped black hole. Its upper reaches are seldom explored or even mentioned [King 1934; Ferris et al. 1936; Briller 1966; Bellet 1971, 1218; Stephenson 1974, 393; Roy et al. 1977; Mandel 1980, 592; see also Fig. 6.3 caption].

By deliberate measurement Shibata et al. [1985ab, 1986] and Ideker and Shibata [1986] have verified the existence of a top to the vulnerable zone in the dog heart, at least for their particular electrode configuration. Their measurements resemble pattern 6.3b with the top of the black hole about 8-fold higher than the bottom in terms of voltage (50- to 100-fold in terms of energy).

The interior of the black hole represents stimuli that transiently reset

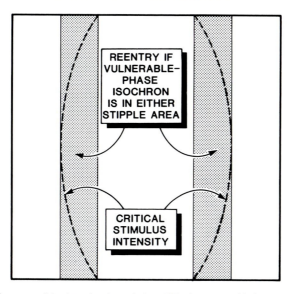

Figure 6.5: If the susceptible tissue has boundaries within the geographical scope of the stimulus, then the even/odd beach may cross the tissue as disjoint arcs (dashed). The critical isochron behind a vertically oriented moving wavefront in Figures 6.1, 6.2, and 6.4 would intersect these arcs only in the stippled areas, during discretely different ranges of time, making the black hole concave in Figure 6.3.

the pattern of timing so as to somewhere include a phase singularity. This provides the seed of arrhythmia. Cell-to-cell coupling may allow it to flourish, initiating rapid reentrant rotations. Changing the choice of stimulus inside the black hole only changes where the centers of rotation appear in the tissue. Changing the stimulus enough to get outside the black hole moves the singularities (centers) outside the susceptible patch of tissue.

There should be such a black hole for depolarizing (excitatory) stimuli if strong enough stimuli evoke even resetting. One might also expect as much without invoking the theory of oscillators, as will be seen in Chapter 7: mirror-image rotors are easily created along the edge of refractoriness behind a wave, by a gradient of premature excitation centered at a point. One might also expect it even without that much elaborate song and dance, simply by considering that a region of nonuniform refractoriness vaguely seems liable to provide reentrant pathways for excitation. This is the more usual interpretation of vulnerability, in which one does not explicitly worry about continuity of electrical state between adjacent cells or the distinction between one-, two-, and three-dimensional wavefronts.

Under the same provisions there should also be a black hole for hyperpolarizing (inhibitory) stimuli. Because in most neural preparations the hyperpolarizing phase singularity lies just before the action-potential

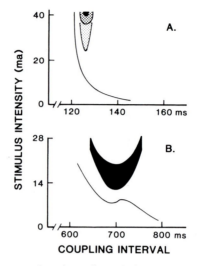

Figure 6.6: (a) In experiments conducted near the end of the refractory period in a cycle of 280 msec, a depolarizing direct current pulse evokes an impulse (extrasystole) in dog ventricular muscle if its intensity lies above the descending threshold curve: less current is needed with later coupling intervals. At higher current, but only within a narrow window of vulnerable coupling intervals, a repeat extrasystole occurs (above the lowest V curve, light stipple). Above a V interior to that (heavier stipple), multiple extrasystoles occur. Following stimuli within a higher V (area shaded black) fibrillation occurs; only the lower tip of this region was measured. Adapted from Matta et al. 1976. The black V of fibrillation is indicated in part (b), redrawn from Brooks et al. 1955, showing a concave top beyond which there is "no response" in turtle ventricle.

upstroke, the vulnerable phase for inhibition would be early in the electrocardiogram's QRS complex, just before the arrival of an activation front. In Box 7.A, forgetting about oscillators and phase resetting, we will also see that rotor pairs are easily created in excitable nonspontaneous media by allowing a transient local inhibition to cut a gap in a wavefront. Such stimuli might include hyperpolarization by neurotransmitters (e.g. acetylcholine), but would probably not include extracellularly applied electrical fields (the stimulus usually intended in discussions of the vulnerable phase). Extracellular currents pass completely through each cell, hyperpolarizing on one side and depolarizing on the other side; the depolarizing effect probably dominates, regardless of current direction. Fibrillation is in fact inducible by either kind of DC electrode, but reportedly more readily by hyperpolarization, for reasons unknown [Brooks et al. 1955; Cranefield et al. 1957; Zipes 1975]. Much remains to be explored with such stimuli in the matter of singularities, black holes, and vulnerability. (See Boxes 6.A and 6.B.)

Box 6.B: Implanted Pacemakers and Cardioverters

Some people use artificial electrical pacemakers to ensure the regularity of their heartbeat. Pacemakers today are much more sophisticated than when the first ones were surgically implanted. Some even have nuclear-powered, radio-programmable on-board microprocessors. But the first ones (called "fixed-rate" or "asynchronous" pacemakers) were mere metronomes, incapable of accommodating a transient spontaneous misfiring or the heart's response to adrenalin, vagal impulses, and the like. A heart reacting to some such extraneous influence could occasionally find itself at an unusual stage of diastole (the relaxation after each contraction) when the next regularly scheduled pacemaker pulse arrives. Ordinarily, this does not cause a problem: the electronic pacemaker fires a fairly weak pulse, seldom exceeding 10 percent of the threshold for vulnerability [Voorhees et al. 1984]. But what if some part of the heart should become 10 times more sensitive than usual? This can happen if some other, benign dysrhythmia starts up and exhausts the tissue, if blood ions get out of balance during shock, if a blood clot or an arterial spasm temporarily cuts off the local oxygen supply, if the autonomic nervous system is briefly hyperactive, or if heavy medication must be given for some other trauma, and so forth. What then? A pulse arriving when the ventricles were almost repolarized after a contraction—in the ventricular vulnerable phase indicated by the electrocardiogram's rising T wave (Figure 2.4)—can induce fibrillation [Surawicz and Zumino 1966]. Such accidents happened with early pacemakers [Tavel and Fisch 1964; Zipes 1975] and have occurred with contemporary implantable cardioverters [Zipes et al. 1984]. It thus became important to measure probability of fibrillation following impulses of all sizes given at all times. The principal result has been that Lown introduced additional timing circuitry to synchronize the firing of artificial pacemakers so that they cannot fire during the vulnerable phase. A byproduct valuable for theoretical understanding was the collection of vulnerability data like Figure 6.6.

Big Black Holes Assembled from
Local Singular Points

What has happened in Figure 6.3? In essence, the depolarizing lower half of Figure 4.9 has been redrawn over and over on the same diagram for many cells whose stimulus coordinates are slightly offset by their different situations. The cloud of singularities has been blackened (then erased along the fringes to mere shading) and the other colors have been left off. In so doing we have fulfilled the anticipation at the outset of Chapter 5 that singularity in a spatially extended oscillator would prove to be quite different from mere uniform quiescence or unpredictable resetting. By attention to the microscopic organization of phase resetting, we have come to the conclusion that some modes of macroscopic arrhythmia may consist of many microscopic repeats of the pinwheel experiment, each producing a vortex of frantic activity around its isolated point of microscopic arrhythmia.

Nonuniformities in the tissue (geographical gradients of timing, stimulation, or sensitivity) have thus enlarged the singular stimulus from a mere point into a finite black hole of vulnerability to reentry, the prelude to fibrillation. Thus phrased, we begin to recognize in these diagrams the classic and widely accepted interpretation of vulnerability (to depolarizing stimuli) as a consequence of spatially disorganized inhomogeneities of voltage:

> When one considers the numerous possibilities by which such a non-homogeneous system [the mammalian heart] might be vulnerable to disorganized behavior, it is indeed surprising that prolonged coordinated activity is at all possible. . . . [M]ost arrhythmias seen in diseased hearts can be produced with relative ease in the normal heart by appropriate activation of the nervous system. . . . [U]nderstanding of the fundamental mechanisms for most disturbances in cardiac rhythm . . . will be based on a clarification of factors which accentuate the non-uniform nature of the heart, and the ways in which such a non-homogeneous system responds to excitation. The time in the cardiac cycle when the atrium or ventricle is most vulnerable to disorganized activity is during repolarization when non-homogeneous electrical properties are maximal. In the atrium, vagal stimulation augments non-uniform recovery and increases the likelihood of multiple responses and fibrillation. In the ventricle, sympathetic stimulation has comparable effects. It is thus well documented that autonomic influences increase the susceptibility of the heart to disorganized behavior in response to an appropriately timed stimulus. [Daggett and Wallace 1966]

Discontinuous dispersion of refractoriness in tissue recovering from depolarization has been widely assumed as the cause of such vulnerability. Nonuniform sensitivity to stimuli and gradients of stimulation may serve much the same purpose. These principles were discussed as early as Mines's original discovery of the vulnerable phase [Mines 1914] and have been much refined and elaborated by generations of physiologists since. By translating them into the language of phase resetting we have only focused attention on the spatial arrangement of that nonuniformity to suggest a specific mechanism for the universally anticipated reentrant activity.

Mother Nature's Dart Board: The Mammalian Heart

Every time you adjust the rate or force of your heartbeat by a volley of nervous impulses, you throw a dart at this bull's-eye. Under normal conditions, they all apparently fall well below it. But conditions are not always normal. When people quip, "I could have died of joy," or "I was almost scared to death," they are referring to real phenomena.

Sudden cardiac death frequently comes unheralded by recognized symptoms of heart disease. People who are presumably at greater risk for some covert reason can and do die unexpectedly—often during an occasion of chronic anxiety and sudden fright or even sudden joy. Cardiologist H.J.J. Wellens, at the University of Limburg, Maastricht, for example, treated a 14-year-old girl who suddenly lost consciousness for the first time upon being awakened one night by a thunderclap. Such syncope is an ominous indicator of circulatory failure in the brain, often indicating ventricular tachycardia or even fibrillation. Fortunately, it seldom lasts long enough for permanent brain damage, but who knows when it might last longer? Wellens's patient would often faint due to tachycardia when startled by her alarm clock in the morning, recovering within one or two minutes (Figure 6.7). A common cause of such difficulties is neural input to the heart along pathways of the sympathetic nervous system, e.g. from the nearby stellate ganglia. Sympathetic nerves activate the membranes of cardiac fibers in complex patterns as the nerve endings ramify across the myocardium. Activation occurs when the nerve ending releases granules of norepinephrine and the cardiac membrane senses them at chemical receptor sites. Those sites, called beta receptors, can be plugged by an artificial beta blocker molecule, usually a nonfunctional, nonremovable chemical analogue of norepinephrine (propranolol, for example). By thus attenuating the impact of sympathetic discharges, the chance of sudden cardiac death is minimized. But the patient must take the medicine. The young girl in Amsterdam did for years, but then she found a boyfriend who may have felt it was ludicrous for fretful old doctors to scare a healthy young girl into chronic anxiety about taking her pills. She discontinued propranolol and was found dead in bed fourteen days later.

Victims of another class are those who have an uncommonly long interval between first ventricular firing and final repolarization. This is called the long QT syndrome, referring to that interval on the EKG (Figure

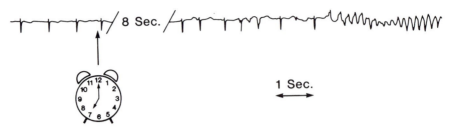

Figure 6.7: Precordial lead VI of a young girl's electrocardiogram shows the onset of ventricular fibrillation about a quarter-minute after she was startled by her alarm clock. During that interval conduction disturbances developed and ventricular premature beats then occurred until one fell in the vulnerable phase. Adapted from Wellens et al. 1972 by permission of the authors and the American Heart Association, Inc.

2.4). In many cases this is a congenital anomaly, inherited along with deafness. Such individuals often perish by sudden cardiac death in re-action to a loud noise or a happy surprise. The untimely triggering stim-ulus is believed to slip through the recovered AV node to strike the ventricles while they still linger in the vulnerable phase. (Normally the ventricles get past that phase while the AV node is still refractory.)

From Sinoatrial Pacemaker Resetting to Ventricular Tachycardia

As Hermann Hesse says in "The Steppenwolf Treatise," "All attempts to make things comprehensible require the medium of theories, mythol-ogies, and lies; and [an author who respects his readers] should not omit, at the close of an exposition, to dissipate these lies as far as may be in his power." The thread of mythology has been stretched rather thin in this attempt to string two-dimensional cardiac reentry on a filament stretching from the space-independent singularities of timing in pace-maker cells (first four chapters) to the vortex-like singularities of rotating waves in three-dimensional chemical media (next three chapters). The thread is already almost invisible in places, perhaps most conspicuously at the transition from electrically uncoupled spontaneously oscillating media to electrically coupled nonspontaneous media. No hidden theorem guarantees safe passage at this point, but a mathematical analogy (see Appendix) suggests that such a guarantee may be discoverable. For the moment it may be preferable to view the matter as follows.

For reasons implicated in that analogy, the phase-resetting behavior of spontaneous oscillators can typically be summarized by an isochronal diagram that resembles a color wheel with grey at the center. This re-setting diagram is measured directly in a pinwheel experiment. Were the pinwheel experiment executed all at once in a spatially distributed way (with a north-south gradient of stimulus size and an east-west gradient of local time, for example) it would produce a wave rotating about the hueless pivot at the period of every local oscillator. Oscillatory media may also be excitable (and neuroelectric media typically are) in the sense that an oscillator near an almost-quiescent phase of its cycle may be hurried past it by a physically adjacent oscillator that is already embarked upon its slightly later phase of spontaneous excitation. In this situation the rotating wave of local timing actually propagates, creating a stable rotor. The rotor's period is short, even as short as the refractory period, and thus generally much shorter than the period of spontaneous local oscillation. It then becomes irrelevant that the local oscillator would have fired spontaneously in isolation: it is always triggered much sooner. A nonspontaneous excitable medium may therefore be expected to behave similarly, once a rotary timing gradient is established.

The mathematical metaphor in the Appendix purports to make plausible that graded stimulation establishes a rotary timing gradient in much the same way whether the local "oscillator" is spontaneous or not. By resort to this metaphor we have organized the successive topics in this book (circadian phase resetting, neuroelectric phase resetting, neuroelectric rotation in excitable media, chemical rotation in excitable media) through the single theme of phase in a cycle, portrayed by color on the color wheel. By evading the more usual technical apparatus of multidimensional state-spaces, invisible trajectories, and unmeasurable manifolds, we thread diverse phenomena on a simply presentable theme of overt timing. But this chapter is truly the thinnest spot on the thread. It thickens again once rotation is established in space: in the following chapters we will use phase arguments and color coding powerfully again to deduce the anatomy of reentrant waves in three dimensions.

If the thread seems too thin here, it is not necessary to trust it. Arguments, after all, are merely aesthetic conveniences for ordering experience. When they grow cumbersome, raw empiricism has greater appeal. Rigorous argument brought us as far as the apparition of a rotating wave elicited by a single spatially graded stimulus falling on a wave in a sheet of independent spontaneous oscillators (the pinwheel experiment). We then accidentally discovered nearby a superficially related phenomenon: a rotating wave elicited by a single spatially graded stimulus falling on a wave in a sheet of coupled excitable cells. As Polya remarks in "How to Solve It": "Good problems and mushrooms of certain kinds have something in common: they grow in clusters. Having found one, you should look around: there is a good chance that there are some more quite near." We can pick mushrooms without digging up the nearby soil to inspect the delicate fungal mycelium from which they all erupt.

Conclusions about the Triggering of Reentrant Arrhythmia

Singularity is almost invariably a clue.
—Sherlock Holmes, in
"The Boscombe Valley Mystery"

Hearts stop for many different reasons, but when they seem to stop for no reason at all, we have a mystery. Physiologists and theoretical biologists have persistently struggled to solve this problem throughout the twentieth century. It is not solved. I have not the slightest doubt it will be, however, before the end of the century; we are that close. But for the present, we have before us only a diversity of intriguing clues, some physiological, some mathematical, some clinical. It is hard to escape the impression that they belong together, that they are the pieces of a single jigsaw puzzle. Maybe a few pieces don't belong; maybe a few are still missing. It is tantalizing.

As far as the experimental, clinical, and theoretical pictures go at present, they seem to overlap agreeably in the idea that fibrillation often begins with reentry, and that reentry stems from singularities of timing, conjured in an appropriate geometrical context. Clinical experience with sudden cardiac death and fibrillation leads to the three broad generalizations that Mines emphasized in his paper about heart muscle:

(1) ventricular tachycardias that forebode fibrillation can be triggered by a premature impulse during a narrow window of vulnerability;

(2) some kinds of fibrillation (probably not all) develop from and entail reentry of circulating impulses; and

(3) fibrillation is favored by almost any departure from spatial uniformity in the target tissue or in its innervation.

We find the same three features linked together by the topological principles involved in resetting a rhythm. But it would still be fair to say that we don't know what fibrillation is or how to prevent it. There are many different ideas about it; maybe there are also many different kinds of fibrillation.

Obviously, this chapter is only an argument, a poor creature of words and qualitative reasoning fashioned to string together some tantalizing facts surrounding sudden cardiac death. If you read the history of science you know the shabby record of such arguments made by biologists. The tools of theory have their limitations. Accordingly this particular piece suffers from inherited deficiencies, notably its emphasis on spontaneous oscillation in a tissue whose rhythmicity is mostly driven or preempted by pulses from a faster pacemaker; and its preoccupation with continuity in a tissue that consists of distinct cells, sometimes scarcely connected, and whose measured behavior seems full of thresholds and discontinu-ities. Indeed, the principal outcome of attempting continuous descriptions is that we are forced to add one more kind of functional discontinuity to the list already familiar to electrophysiologists: the phase singularity. Other theories of fibrillation have their own sore spots: the classical allusion to dispersed refractoriness treats the tissue as though it were a three-dimensional fish net of independent one-dimensional conduction paths; the "wandering wavelets" idea has difficulty with the demonstrable regularity of electrical activity during fibrillation; the "choral mode" idea underplays the observed residue of vorticity in activation patterns; the notion of overlapping waves from multiple foci of rapid spontaneous activity neglects to account for the simultaneous appearance of those foci and their similar extremely short periods. Like other theories of fibril-lation, this one with wavefronts connecting labile phase singularities will mature into a quantitative description or else be quietly forgotten only when we can assign numbers to the processes dominant when time breaks down in cardiac muscle:

when we can compute phase resetting accurately in each of the five distinct tissues of the human heart, both when spontaneous and when merely driven periodically, to know the coordinates of the two phase singularities in each: do any coincide with the known vulnerable phases? (see Figure 6.8);

when we know the geographical distribution of sympathetic and par-asympathetic neurotransmitters during neural regulation of the heartbeat, and know their effects on timing quantitatively; numerous studies even twenty years old show that their effects are local (but how local?), graded (but what maximum impact?), and often initiate arrhythmias presumed to involve reentry (but exactly how?);

when we can formulate numerical criteria for the stability of rotating waves in excitable media, and relate them to the measurable properties of cardiac muscle;

when we can realistically incorporate into our thinking the myocar-dium's actual anisotropies (of conduction velocity, e.g.) and organic in-homogeneities (of repolarization rate, e.g.), both normal and due to in-farcts. They exist. They undoubtedly provide other mechanisms for creation of singularity pairs, and other mechanisms of arrhythmia. (For example, see Spach et al. 1981; Spach et al. 1982; Spach and Kootsey 1983.)

These are currently subjects of energetic investigation by dedicated physiologists in every technologically advanced country. But it will still be a few years before alternative visions of malignant arrhythmias—or

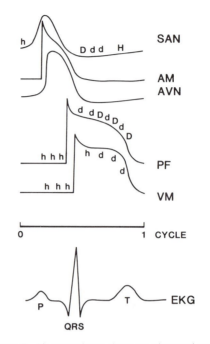

Figure 6.8: Figure 2.2 is copied with approximate times superimposed on each trace when laboratory measurements or calculations from electrophysiological models show a phase singularity in response to hyperpolarization (H for data, h for computations) or depolarization (D for data, d for computations) of appropriate strength. For SAN, experimental data are from Jalife and Antzelevitch 1979; computations are from Reiner and Antzelevitch 1985. For PF, experimental data are from Jalife and Antzelevitch 1980 and Antzelevitch and Moe 1983; computations are from Chay and Lee 1984. For VM, computations are from Chay and Lee 1985. Because available data use VM and PF preparations made rhythmic by a constant biasing current, h is poorly located just before the action potential. Neither experiments nor models are yet available for AM or AV. Only in Purkinje fibers and ventricular muscle (understandably!) does the depolarizing singularity come close to the observed vulnerable period for ventricular fibrillation.

at least of their onset in patterns of reentry—can confront quantitative facts. Maybe they will all prove correct, each under its own circumstances. Maybe they are all dreams and our awakening still lies in the future.

This is as far as theory takes us for the present. It has brought several clues together in a suggestive way. But are we any better off for it? Is our situation only as reported by the industrial consultant cited apocryphally at the end of Chapter 1? Suppose it is true that sudden cardiac death often begins with singularities of timing surrounded by little vortices of electrical activity. Does it do any good to know this? If the initial singularities are indeed seeded by a single nervous impulse falling precisely in the critical phase, there is probably not a lot to do about it. The human heart is continually bombarded by impulses of obscure and multifarious origin. They might be regarded as darts raining onto a target whose bull's-eye—the black hole—like the iris of a camera, can be enlarged or constricted but cannot be removed. The job of natural selection (and preventive medicine) is then to make the black hole as small, inaccessible, slippery, and repellent as possible, perhaps deepening its fringe of grey by arranging that adjacent reentries promptly fall into one another's arms after emitting only a few ectopic beats. Natural selection has done this job so well that fibrillation is an uncommon cause of death in hearts unstressed by coronary artery disease. If evolution has ever achieved an engineering triumph, it has been achieved in the electrical control of the mammalian heart, engineered for tough, fail-safe performance without a lapse. Its persistence is remarkably uniform from hearts as tiny and vibrant as a shrew's to those as ponderously slow as a whale's: any mammal's life spans about 2 billion beats [McMahon and Bonner 1983]. Even then, the heart is seldom the ultimate cause of death; fibrillation comes only if bidden by some other condition that makes the heart more than usually vulnerble. A machine engineered this well is not to be approached overconfidently with glib mathematical "explanations," but rather with the words of Albert Einstein (1922) in mind:

> The scientific theorist is not to be envied. For Nature, or more precisely, experiment, is an inexorable and not very friendly judge of his work. It never says "Yes" to a theory. In the most favorable cases it says "Maybe," and in the great majority of cases simply "No." If an experiment agrees with a theory it means for the latter "Maybe," and if it does not agree it means "No." Probably every theory will someday experience its "No"—most theories, soon after conception.

The crux of the challenge for preventive medicine is concealed in the metaphors of the previous page: "small, inaccessible, slippery, repellent." Throughout this chapter the question of stability was evaded. This is where the secret still lies hidden: does a phase singularity turn into a rotor? Does a rotor pair recombine into spatially uniform activity, or does it metastasize into many scattered rotors? If a single nerve impulse can end one's life so easily, why doesn't it happen sooner? Presumably all hearts are subject to the same erratic impulses, but the seeds of turbulence flourish only in hearts somehow predisposed to electrical instability. How predisposed? What makes the soil fertile for rotating wavelets? How can we identify the population at risk? Clearly, further improvements of preventive medicine are best sought within the context of clinical and epidemiological medicine. But sudden cardiac death may be one of those exceptional phenomena in physiology that invite meaningful contribution from applied mathematics. If fibrillation or its preamble, reentry, depends on the stability of rotating waves, then it would be well to understand such waves. In particular, we must understand how

Box 6.C: A Rationale for Prescription

In the normal activation of human ventricular muscle, the whole area of the endocardial (inside) surface is almost synchronously ignited by the Purkinje fiber arborization. Excitation proceeds to the epicardial (outside) surface through myriad parallel pathways. In contrast, during abnormal self-activation, the impulse actually circulates transversely to the epi- and endocardial surfaces. Why is this mode more stable in some hearts than in others? First of all, it is stable in all human hearts, given appropriate initial conditions (e.g. electrical shock), but in some hearts those initial conditions are more readily achieved.

One approach to limiting the incidence of rotary excitations is to modify the innervation of the heart, e.g. by severing the sympathetic innervation from the stellate ganglia. Another approach entails pharmacological (e.g. propranolol) neutralization of the receptor sites in the cell membrane so that neurotransmitters cannot affect electrical performance. Yet another approach entails alteration of the electrical properties of the cell membrane itself. But in what way? A common idea is that the propagation speed should be altered (by changing the speed of initial depolarization) and the refractory period altered so that a reentrant wave would not have room enough for stable circulation. There are many other ideas too, as indeed there should be, considering that there are diverse mechanisms of instability to be guarded against, some more urgently than others. Krinsky [1978] reviews some of these, emphasizing reentry.

The more recent publications of Zykov, summarized in his 1984 book [Zykov 1984, in Russian; for English see Kogan et al. 1980], also attempt to create a rational quantitative basis for prescription by numerical solution of the partial differential equations of activation. Zykov argues that the two most important membrane parameters are pulse duration and the initial steepness of depolarization. Each kind of anti-arrhythmic drug alters both in characteristic proportions and directions. On a plane defined by those two parameters, the action of the drug is to move the membrane in a characteristic direction. What direction is wanted? Zykov suggests the direction in which the minimum radius of reentry is larger. He calculates the minimum radius at each point in this plane and fills the plane with contour lines of uniform radius: safety lies uphill on this contour map. Surprisingly, the contour map resembles a deep valley: which way is uphill depends on where you start. Depending on the present state of the patient, this direction may be as is traditionally assumed from looser analyses or may be quite different, even opposite. The two slopes of the valley represent stationary circulation (where conventional notions of how to prescribe against reentry are exactly counterproductive), and meandering circulation (where intuition seems correct).

Experimental confirmation of Zykov's picture could have great practical value.

such waves are affected by changes in membrane properties that accompany ischemia, excessive release of neurotransmitters, and medication.

There are several classes of anti-arrhythmic drugs in current use. Each class has its own effects on the various excitability properties of the heart's several distinctive tissues. Some of those properties are more involved in dysrhythmias resulting from abnormal impulse initiation (ectopic pacemakers). Others more affect dysrhythmias arising from abnormal conduction: pulses circulating on a one-dimensional closed path ("ring" reentry: the classical "circus wave"), or rotors in two dimensions ("vortex reentry": here and Chapter 7) or in three dimensions ("organizing center reentry": Chapter 8). We know enough to distinguish ring reentry from vortex reentry clinically by their differential responses to lidocaine, for example [Agladze et al. 1983; Grenader and Zurabishvily 1984]. But we still lack a thorough understanding of the impact of drugs on wave propagation and stability (see Box 6.C). In the absence of any tested and widely accepted model of rotor stability, medical prescription remains an empirical art, and some of its successes seem counterintuitive. Enthusiasm over protective agents such as bretylium, for example, is tempered by uncertainty about their mechanisms of protection. Cardiologists are productively busy about many approaches to the many modes of sudden cardiac death, chemically and surgically picking the locks of many doors. A rich harvest of improvements may also be waiting behind a door whose combination lock is a partial differential equation.

Such equations can be as hard to understand as the original medical problem. During the past decade, biophysicists have been increasingly attracted to a supplementary approach that combines computer experiments with laboratory investigations of chemical analogues of biological clocks and excitable media. The results illuminate mechanisms of oscillation, of phase resetting, of singularities and arrhythmia, of wave propagation and in particular, of rotating waves. Cardiac dysrhythmias will only be deciphered by experiments in electrophysiology and cardiology, and these will take time. Meanwhile, similar phenomena are almost understandable in a more convenient experimental system, an oscillating and excitable medium powered by known chemical reactions: see Chapter 7.

Chemical Rotors

> Undoubtedly there should also be a real and general
> biology but we can only just begin to glimpse it. A
> true biology in its full sense would be the study of the
> nature and activity of all organized objects wherever
> they were to be found.
> —J. D. Bernal (1965)

The internal organization of a bioelectric vortex might be compared with what is found in the core of a tornado: unfamiliar states of the medium, steeply graded in ways that cannot relax due to the dynamics of circulation outside. Admittedly, neither is yet terribly well understood, in both cases for the very reason that requires us to understand them sooner or later: they are hard to confine stably, and they inflict violent death abruptly and unpredictably, then vanish.

It is difficult to play with such things in the laboratory. We still know very little of rare diseases, earthquakes, lightning strikes, ball lightning, volcanic eruptions, sudden cardiac death, stroke, crib death, or tornados because no one has yet found a conveniently simple and reproducible instance to experiment with. Wouldn't it be nice if there were some convenient fluid, something like "extract of excitability" or "biological clock juice" that lent itself to re-creation of biochemical whirlwinds in their starkest, simplest form for study? Yet there is. It has been known within a tiny circle of cognoscenti since about 1950 but was hushed up as a subversive affront to theory until it was fourteen years old. There were, of course, earlier encounters with other oscillating reactions [Morgan 1916; Bray 1921; for an update see Field and Burger 1985], but they were conceptually inconvenient and were ignored until theorists wrote their licence to exist. There were also much earlier encounters with chemical excitability by itself, divorced from spontaneous oscillation. Since the beginning they have been studied with the neuroelectric analogy in mind.

Chemical Waves

The chemist William Ostwald was the first to describe nerve-like "excitability" in a chemical reaction [Ostwald 1900]. With a zinc needle Ostwald pricked the dark coating of oxidation that covers an iron wire bathed in acid: it changed color, creating two electrochemical pulses that scurried away at speeds approaching a meter per second in both directions. This traveling pulse of darkness behaves in every way like the invisible action potential initiated by exciting the membranous surface of a long nerve axon: there is a definite threshold below which a stimulus will not trigger a propagating pulse; a stimulus transgressing that threshold elicits a pulse of characteristic shape that cruises without attenuation at a characteristic speed; two colliding pulses annihilate one another; the pulse vanishes without reflection when it strikes a barrier. As Ryoji Suzuki first showed me, a pulse can even be induced to circulate forever on a big enough loop of iron and silver; we seemed to be watching a living organism while these pulses leapt and darted at lightning speed like silvery minnows along the submerged wires.

For a while many biophysicists hoped to understand nerves by first understanding the simpler chemical analogue. There arose a substantial literature of experiments and theory, largely at the hands of Ralph S. Lillie at the University of Chicago during the 1920s and Karl Friedrich Bonhoeffer in Germany in the 1940s [reviews by Suzuki 1976, MacGregor and Lewis 1977]. In the 1950s the work of K. C. Cole in the United States and A. L. Hodgkin and A. F. Huxley in England [Cole 1968] made nerve membrane itself much more tractable for direct study.

During the 1960s, however, many scientists still wanted to study propagation in a two-dimensional context, and the iron/acid/silver analogue remained the only practical alternative to pure computation or mathematics. E. E. Smith and A. C. Guyton in the University Medical Center of Jackson, Mississippi, filmed a ten-inch sphere of iron suspended in a bath of nitric acid [Smith and Guyton 1961]. Stimulated at a point, this "iron heart" conducted a visible pulse of chemical change in an expanding circle from the initial point to annihilation at the antipodes. In hopes of detecting the rotating waves long believed responsible for the most pernicious arrhythmias of heart muscle, Smith and Guyton then provided a burst of short-period stimuli. It provoked waves that maintained themselves in complicated swirling patterns long after the stimulus was removed. But bubbling and convection in the acid bath made serious study virtually impossible.

Meanwhile in Tokyo, Jin-Ichi Nagumo and students filmed the waves of darkness on a technically more convenient flat gridwork of interwoven iron wires [Suzuki et al. 1963]. This ingenious experiment, published only in Japanese until 1976 [Suzuki 1976], went unnoticed in the West.

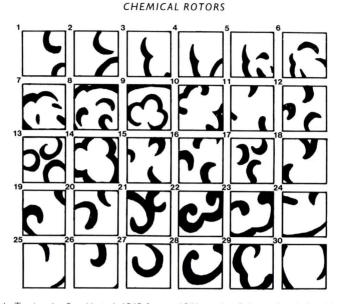

Figure 7.1: Tracings by Suzuki et al. 1963 from a 1961 movie of electrochemical activity on a 26-wire × 26-wire grid of iron immersed in nitric acid: thirty frames at intervals of 1/8 second from left to right, then next line from left to right, etc. This was the first experiment, as far as I know, to demonstrate waves pirouetting on a surface without holes.

Figure 7.1 is a pencil sketch from the original publication, perhaps the first to show fragmented waves rotating uncertainly about volatile pivot points, much as in recent electrical recordings from heart muscle during stable atrial fibrillation (Figure 7.2, from Allessie et al. 1985). Both illustrations are sketchy and seem to contain internal inconsistencies, perhaps no surprise in converting the interplay of several continuous variables to black-and-white discretized snapshots. Though they do not bear close scrutiny, they hint at vortex-like motion and encourage interest in analyzing more detailed representations.

By the time of Nagumo's experiments, chemists in the Soviet Union were developing a much more convenient chemical incarnation of the principle of excitability: the Belousov-Zhabotinsky reagent.[1] The discovery was essentially complete in 1951, but it was written only in Russian, the emphasis was on oscillation (sweeping excitability under a rug), spatial aspects were neglected, and it was denied publication anyway. All this was abundantly corrected by 1980, when Boris Belousov, Albert

---

[1] Diverse names have been used, not always consistently, e.g. Zhabotinsky reagent or malonic acid reagent (in which Zhabotinsky replaced Belousov's original citrate by malonate), or Z-reagent (in which by a change of recipe proportions and exposure to air in a thin layer I suppressed the malonic acid reagent's capacity for spontaneous oscillation). "Belousov-Zhabotinsky reagent" generally refers to all of them.

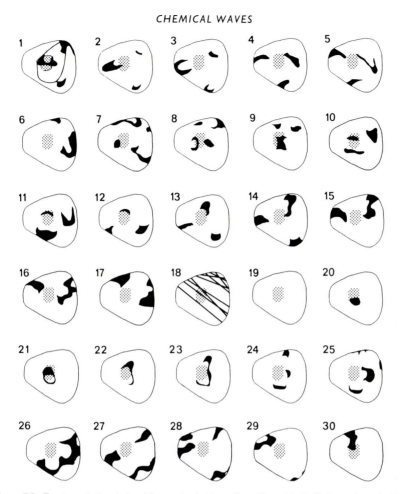

Figure 7.2: Tracings of electrical activity on the inside surface of a stably fibrillating atrium of a dog: thirty frames at intervals of 1/100 second from left to right, then next line from left to right, etc. The endpoints of shaded activation zones here seem sometimes to pivot, sometimes to move with the wavefront; such vagaries may stem in part from the difficulty of consistently deciding in each electrode channel what is an activation signal and what is too attenuated to qualify. In panels 18 and 19 all activity ostensibly vanishes, to erupt again from a central spot in panel 20. I use panel 18 to show how the image of the closed surface of the atrium was cut open for this flat presentation: lines link adjacent points along the incision. Electrical activity on either edge should appear at the corresponding point on the other edge and often does in these sketches (the links are taken from panels where it does). The superimposed outline of the uncut atrium in panel 1 shows an inexcitable region where the superior vena cava enters, here replicated onto each panel. The sketched activity often seems peculiar near this region. Adapted from Allessie et al. 1985 by permission of Grune & Stratton, Inc., and the authors.

Zaikin, Anatoly Zhabotinsky, Valentin Krinsky, and Genrik Ivanitsky received the Lenin Prize for work on this excitable medium. We will examine this discovery below. But first, why do we care?

Expanding Our Mandate

We care because little in science is well deciphered by attention to any very limited class of phenomena. As the physiologist A. G. Mayer groused [1906], "The papers of physiologists abound in general conclusions concerning the action of 'the vertebrate heart' when only the heart of the terrapin or the dog has been studied, and undoubtedly these sweeping conclusions are often misleading."

Many details of a particular phenomenon's mechanism may be understood, but without a supportive context of related processes, it is easy to overlook the essential principles, indistinguishably lost among the particular details. By struggling with two or three related phenomena at the same time and trying to appreciate each from the perspective of the others, we suffer certain hardships but also enjoy opportunities to notice common themes and perhaps come to a simpler understanding of the essentials. No one really understands the mechanisms of reentry in two- and three-dimensional media, physiological or otherwise; and in physiological situations, the basics are obscured by anisotropies, inhomogeneities, discontinuities, and fluctuations. Simpler excitable media exhibit one-dimensional wave phenomena remarkably similar to those seen in, for example, heart muscle. Wouldn't it be surprising if their two- and three-dimensional wave phenomena do not also appear, perhaps with baroque modifications, in living media whose complexities make initial perception more difficult?

Let us therefore review the present unsatisfactory state of our attempts to integrate spatial and temporal thinking about phase singularities and then see what a new clock-like and excitable medium has to teach us.

Rhythmicity and Excitability in Spatial Context

In the Introduction, attention was invited to the biggest known biological clock: the Earth's living skin, its biosphere, sedately rotating through 24-hour cycles of light and darkness. The underlying continents and oceans periodically roll into the sunlight and back into shadow not because each "has an internal clock" but because each has neighbors pushing from behind and pulling from in front. Similarly, reentry about an arrhythmic pivot can arise in muscle despite the fact that its constituent fibers (like the rocks and trees composing the rotating Earth) need not individually possess any innate rhythmicity. Rather, each fiber is simply compelled to follow its neighbors, just as the Earth's rocks must, in remorseless rotation once the cycle is closed.

This is a critical shift of perspective. In Part I we discovered the susceptibility of biological clocks to the triggering of instant arrhythmia. This phaseless state is often only as stable as the stimulus is exact, and therefore it quickly converts to a randomly phased rhythm. This lability is fixed by grafting the concept of a singularity of timing into the context of Part II: embedding it in two-dimensional space as a pattern singularity of space and time intertwined. In Part I our charter was limited to discussion of "clocks" but in Part II it was amended to permit discussion of cells incapable of spontaneous rhythmic firing when isolated, just as long as they do fire periodically anyway. Treating the whole heart as an oscillator, we found that this spatially structured clock's phase singularity is a spatially structured alternative to the normal cycle: in one manifestation, it is a wave rotating within the sinus node, not simply a uniform arrhythmia. Similar waves are found even outside the spontaneously periodic tissue. Thus, by incorporating physical space into our original purely temporal charter, we could dispense with strictly spontaneous rhythmicity. Some attempt at justification is contained in the Appendix.

Another amendment is now overdue. The notion of continuity originally served as the bridge from observation to the next prediction. Specifically, we emphasized continuity in the dependence of an oscillator's phase resetting on the parameters of an externally applied stimulus. But one kind of continuity might be substituted for another. A substitution is needed because in the previous two chapters and now in the chemical context of the remaining chapters, we deal with excitable media in which a trigger provokes thousandfold changes of chemical activity in mere thousandths of a typical cycle period. In that respect chemically excitable media resemble bioelectric membranes harboring populations of fast sodium channels. Strictly speaking, even such quick-changing population averages follow continuous kinetic rate laws, but there is little to gain practically by insisting on the point. (And still more strictly, membrane events are discrete, molecular, and stochastic: see Box 3.A.) It will be more convenient to shift our emphasis to preoccupation with continuity in the spatial arrangement of timing. We may infer some principles of timing, of the organization of timing in space, and of the breakdown of timing at unavoidable singular foci from the requirement of spatial continuity alone.

Earthquake Fault Lines of Timing

As noted in the previous chapter, influences at work on rhythmically active media often create "cracks" in the spatial pattern of timing by resetting a block of the medium quite differently from an adjacent block. Nearby places thus sometimes have very different phases. If adjacent places fire independently without triggering their neighbors, then a phase discontinuity means firing spontaneously at considerably different times. A wave that progresses smoothly across a region of smoothly graded phase will appear to creep across the earthquake fault line of local phase, almost standing still, before leaping forward again from the far side of the crack at an apparent "speed" inversely proportional to the phase gradient. But persistent discontinuity requires complete independence of kinetics in adjacent regions. In a chemically reacting liquid, molecular exchange goes on between places very close together. Without it, a parochially oscillating liquid would gradually become incoherent, since the clock could not be expected to run at exactly the same pace everywhere. And a nonoscillating excitable liquid would do its parochial thing just once and then revert to uniform quiescence. Things are quite different when exchange between neighbors occurs: oscillations remain synchronous, waves propagate, and excitation circulates where peace might otherwise reign. It is because this coupling assists neighbors to skip over the lingeringly slow parts of their local kinetics that rotors in chemical media, just as in neuroelectric media, spin several times faster than the same medium would oscillate uniformly (if at all).

Periodic Activity in a Chemically Reacting Liquid

The first chemical oscillator to command worldwide attention was discovered by Boris Pavlovich Belousov [Winfree 1984a; Zhabotinsky 1985]. Belousov was a biochemist at Moscow State University; he was already 57 at the time of his most remembered discovery. Like most biochemists of that day, he was keenly aware of the wonderful sleuthing job that would win the Nobel Prize for Hans Adolf Krebs of Oxford two years later, in 1953. Krebs had found out how organic carbon chains are finally burned to carbon dioxide while their energy of oxidation is captured in molecules of NADH and ATP. In this process the di-carbon victim is welded to an organic acid to form citric acid, which is then indirectly converted back to the original acid by chopping off two carbon dioxide molecules. This recycling is, naturally enough, called the Krebs cycle or the citric acid cycle—not because it is rhythmic in time but because the reaction steps lead in a circle. Belousov undertook to create an inorganic caricature of the citric acid cycle by oxidizing citric acid, not with catalytic enzymes, but merely with the metallic ions that enzymes commonly carry in their active sites. He chose cerium, giving the solution a faintly yellowish tinge, and he replaced the oxidizing enzyme cofactor NAD by inorganic bromate in a solution of sulfuric acid.

Much to Belousov's surprise, his vial of citric acid solution actually cycled rhythmically in time: in about a minute the yellow of oxidized cerium ions faded while they in turn oxidized the citric acid, and then returned in the next minute as bromate ions oxidized the reduced cerium!

In 1951 Belousov submitted to the appropriate Russian chemical journal a manuscript entitled "A Periodic Reaction and its Mechanism." Unfortunately, it was believed at the time that a solution of chemical reagents could not oscillate: the most superficial understanding of thermodynamics guaranteed that such reactions must proceed directly to their final state. *Only* the most superficial understanding can "guarantee" such monotony, but the referees of Belousov's manuscript did not make such subtle distinctions. They rejected his experimental evidence as obvious fraud or the fruit of gross incompetence.

Undeterred, Belousov worked six more years to elucidate in some detail the chemical mechanisms of his astonishing observations. The finished report of 1957 was sent to a different chemical journal. It, too, was rejected, this time with a cutting remark about "the supposedly discovered discovery" and a requirement that before publication of his experiments could be considered, Belousov must demonstrate to the satisfaction of theorists the error of their concepts. This time Belousov the experimentalist was embittered and withdrew, resolving never to share this discovery with his erstwhile "colleagues" (Figure 7.3).

It is easy to be unsympathetic with the anonymous referees and it is difficult to step back into the attitudes prevailing among scholars at that time. Even in the summer of 1968, when I first heard of Belousov's reaction from fellow graduate student Zhabotinsky in Prague, my feelings fluctuated between amazement and skepticism. Months later in Princeton with Zhabotinsky's recipe in hand, I slipped a cuvette of the mixture into a recording spectrophotometer. The pen swept left as the liquid darkened in the invisible ultraviolet bands—slowed—paused—and returned. The cuvette was clearing! But then the clearing slowed—paused—and darkened again toward extinction, only to slow again. The phone rang; when I returned, there was the rhythm of optical density in the invisible ultraviolet region (Figure 7.4).

As a test of what people were prepared to believe, I showed this "problem" to colleagues in charge of the chemical analysis lab: they sought every possible electronic and mechanical "bug" in the recording machine before contemplating the ludicrous possibility that the reaction observed was actually oscillating!

To return to Russia in the late 1950s: Belousov's recipe was circulating among chemists in the Moscow area. It aroused great interest but Belousov never published more than a brief abstract, unknown to his colleagues until years later. It was without illustrations and in a Russian-

Figure 7.3: Boris Pavlovich Belousov about 1956–1958. Courtesy of Professor S. E. Schnoll. From Winfree 1984a; used with permission of the *Journal of Chemical Education*.

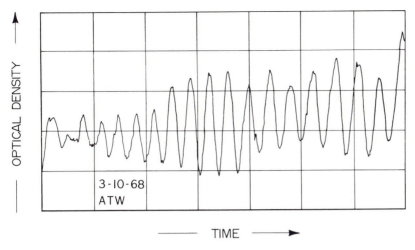

Figure 7.4: Optical density in the ultraviolet, looking through 1-cm thickness of unstirred, bubbling Belousov-Zhabotinsky reagent. Scale divisions are five minutes apart. From Winfree 1980.

language symposium proceedings on a different subject [Belousov 1959].

Real progress in the investigation of this "liquid clock" resumed in 1961 when S. E. Schnoll assigned graduate student Zhabotinsky to find out what this apochryphal recipe really does. Zhabotinsky replaced citrate by malonate and refined many other details. The 1960s then witnessed a flurry of publications about this oscillating chemical reaction (and others, e.g. glycolysis in yeast cells; starting in the mid-1960s and continuing still, the annual rate of publication on oscillating reactions has doubled every three years, five times the doubling rate for chemistry as a whole). By 1970, all that Belousov had tried to report had been rediscovered in a considerably more exact and refined way in the Soviet Union, Denmark, Belgium, Germany, and the United States—though his name remained virtually unknown outside his own country. Belousov died in 1970; his Lenin Prize was awarded posthumously in 1980.

Chemical Waves Revisited

In 1970 two astonishing new discoveries came to light: that this chemical medium supports waves of chemical activity and that those waves emerge periodically from tiny regions of chemical rotation. These are the subjects of this chapter and Part III. Before focusing on these aspects of its behavior, it will be useful to examine this new chemical medium in relation to the topics of Parts I (oscillations and their critical annihilating stimuli) and II (the same in spatial context: excitability and reentrant propagation).

The differences, to begin with, are obvious enough. The chemical solution is homogeneous, continuous, and isotropic, whereas heart muscle, to take an extreme case from our previous topics, is none of these: it is inhomogeneous in more ways and on more scales of time and space than could be listed on an entire page and the cells are fibers whose orientation (and preferred conduction direction) varies systematically. Such differences cannot be casually overlooked (see Spach et al. 1981, Spach et al. 1982, Spach and Kootsey 1983, and Diaz et al. 1984). These differences are, however, the very reason for attention to the residual similarities: it is essential to find diverse excitable and clock-like media in order to distinguish which of many behavioral capacities derive specifically from particular mechanisms and anatomical idiosyncrasies and which are only modified thereby, being inherent even in perfectly idealized caricatures. We now turn to similarities between the chemical reagent and (collectively) the objects of our earlier attention.

First of all, it is clear that Belousov's reagent constitutes a liquid chemical clock; after all, that was the original reason for its suppression, for its explosive breakthrough to international attention, and in particular for my importing the malonic acid version from Zhabotinsky in the summer of 1968 into my biological clocks laboratory. Like any other attracting-cycle oscillator, it is susceptible to phase resetting by a discrete impulse, e.g., of ultraviolet light or an injection of a tiny amount of a suitable chemical. It shows odd-type resetting in response to small stimuli and even-type in response to stronger ones [Sevcikova et al. 1982; Dolnik et al. 1984; Ruoff 1984].[2] Assembled three-dimensionally, detailed resetting measurements present a time crystal with helicoidal surface and singularity just as in Chapter 1 [Dolnik et al. 1986]. Like circadian clocks, cardiac pacemakers, or any other limit-cycle oscillator, the reaction can be entrained by periodic perturbations [Dolnik et al. 1984].

Second, the malonic acid reagent is excitable. During each cycle there is a prolonged, relatively quiescent interval when nothing seems to be happening. Actually, the reaction is then lingering quite near to a steady state in which rates of synthesis and degradation of each molecular species are nearly in balance. In the spontaneously oscillating version, they are not quite in balance. In particular, bromide ions are slowly sticking onto organic acceptors, and when too few remain free, a reaction will be

---

[2] Other chemical oscillators (and perhaps this one too, perturbed in a different way) exhibit apparent discontinuities in their phase-resetting behavior, e.g. the Briggs-Rauscher reaction studied by Dulos and De Kepper [1983]. In the cited instance this is apparently due to rapid kinetics, as a slightly different recipe gives smoother oscillations and smooth even-type resetting [Dulos 1982]. In other instances it may be a consequence of measuring only the first phase-marking event after the perturbation [Kawato 1981]. See discussion of discontinuities in Chapter 3.

disinhibited that suddenly changes everything and restarts a new interval of quiescence. Is the excitability of malonic acid reagent dependent on its spontaneous cycling? Not at all: by fiddling with the recipe, one easily converts the "lingering" near-steady state into a stable balance (see Box 7.A, below). But if that balance is upset by a sufficient disturbance away from the steady state, the explosive stage of the reaction is initiated, as usual. Just as in touching a spark to a keg of gunpowder, a single cycle of chemical transformation results, ending in quiescence again: but in this intriguing case, that quiescent state is not a state of inexcitable exhaustion. Rather, it is a state identical to the first except for the negligible loss of about one percent of the reactable ingredients. Perhaps a better analogy is to a grass fire: a lightning strike ignites the accumulated tinder of many growing seasons, leaving almost-quiescent ashes, roots, and maybe seeds; they grow, restoring the grass and, in due course, the tinder. The prairie is then ready for another grass fire.

A still better analogy—remarkably close, in fact—was encountered in propagating action potentials (Chapters 5 and 6). Nerve and muscle membranes rest near an electrical steady state determined by the openness of protein pores through which specific metal ions can slowly enter. Anything that transiently enhances this influx diminishes the electric potential, which allows the pores to open further. If the upset is big enough it initiates a much bigger upset, leading, however, eventually back to the original quiescent state. As in the malonic acid reagent, in some versions of nerve and muscle membrane this quiescence is only a lingering passage near the true steady state, periodically erupting into a spontaneous action potential. One such version is the pacemaker node of the mammalian heart.

Third, just as in biomembranes, the malonic acid reagent contains several crucial ingredients reacting against one another. They take quite distinct time-courses, one rising explosively then decaying exponentially, one rising and falling in an abrupt pulse, another rising to a plateau from which it comes down only after other reactions have finished.

Fourth, and most important, adjacent bits of membrane "talk to one another" in much the same way that adjacent droplets of the chemical analogue do. In the malonic acid reagent, the exchange involves molecules diffusing from regions of higher concentration to regions of lower concentration—e.g. from a place where they were recently synthesized more quickly than they were degraded, to an adjacent place where that imbalance hasn't yet been triggered (but the exchange may trigger it). In this exchange the submicroscopic life histories of individual indistinguishable molecules are not as important as the statistics of their populations: the concentration or "chemical potential" changes according to the mathematical laws of diffusion, the same rules that govern heat flow from hot to cold. They also govern the flow of electric potential from high voltage to lower. In the case of cell membranes, this is what diffuses to couple adjacent regions: electric potential, carried by currents of charged ions (Figure 2.3). The coupling is mathematically the same,[3] and so is the consequence: excitation propagates, just as it does in a grass fire.

Propagation implies direction. The direction is determined by history: one region fired first, turning to inexcitable ashes temporarily, and its excitable neighbor fired later, infected from the first. This asymmetry (excitable quiescence ahead, refractory ashes behind) gives a direction to the wave so that it cannot backfire. Thus when two waves collide, they should snuff one another out: neither can propagate into the ashes left behind the other. Just as though a wave were colliding with a virtual mirror image of itself, the same annihilation occurs at any barrier, be it a firebreak, dentist's novocaine surrounding a segment of cranial nerve, or the glass wall of a chemist's container: waves of this kind should not reflect from barriers, as there is nothing behind to burn.

Belousov may have been the first to see propagating waves in this reagent, though it is not clear that he recognized them as such; no mention of them appeared in the unpublished manuscript that he gave Zhabotinsky in 1962. In the 1981 archival version, however, he thanks his colleague A. P. Safronov for suggesting the use of ferroin dye to enhance the visibility of regional variations of chemical activity and mentions wave patterns like geological strata in an unstirred cylinder of reagent. In any case, the first published account of such striations came from Heinrich Busse, then a student in Germany [Busse 1969]. Busse had independently discovered the ferroin trick but, beguiled by a theoretical anticipation of stationary ripples in the concentration pattern, did not clearly appreciate the blue bands as moving fronts of excitation on the quiescent red background. A clear formulation in these terms was left to Zaikin and Zhabotinsky [Zaikin and Zhabotinsky 1970].

Their report appeared in English in early 1970, together with photographs like those in Figure 7.5. Here we see waves emitted from time to time by mysterious points analogous to ectopic foci on the heart. At those irritable points (adventitious chemically active dust motes in the liquid, perhaps) the chemical clock ticks faster. So those points turn blue (light grey in Figure 7.5) before their surroundings, and the red-to-blue (dark-to-light) transition propagates (until the whole region ahead turns blue [light] spontaneously) in concentric rings like ripples from a stone thrown

---

[3] Almost the same. In membranes, only the electric potential diffuses, not the several other state variables that concern the average proportion of open protein channels for specific ions. In the chemical case, all reactants diffuse, hydrogen two or three times faster than most other small ions and molecules.

Figure 7.5: Four snapshots taken 1 minute apart, and a fifth somewhat later, showing blue (lighter grey as reproduced here) waves of oxidation in a 66-mm dish of orange (darker grey) malonic acid reagent. Waves erupt periodically from about ten pacemaker nuclei scattered about the dish. In the end only those of shortest period survive. The tiny circles are growing bubbles of carbon dioxide.

Figure 7.6: As in Figure 7.5, but instead of concentric rings we have two radially expanding mirror-image spirals of oxidative activity. The small circles are bubbles of a reaction byproduct, carbon dioxide. The fine white streaks are scratches in the 16-year-old negative, which was made by ATW with the support of the Medical Research Council Laboratory of Molecular Biology, Cambridge, England.

in a pond. But unlike ripples, which would pass through those from another stone, these waves of chemical activity do indeed mutually annihilate in head-on collisions with waves from adjacent pacemaker points. Such waves more resemble grass fires or epidemics recurrently spreading into susceptible territory than the interpenetrating waves of familiar physics. As fire and epidemic analogies would suggest, these waves do in fact vanish without reflection at the edges of the dish.

These are not waves of liquid movement but waves of chemical change, like action potentials succeeding one another on a nerve axon. Like action

Figure 7.7: As in Figure 7.6, but many spirals coexist.

potentials, their onset is so abrupt as to constitute a virtual discontinuity: the wavefront spans only a few percent of a millimeter; during mere milliseconds, one of the reactants explosively increases a thousandfold in concentration. Unlike water waves or sound or light, waves of this kind propagate steadily without attenuation. Though the wave may be spreading from a point source, it does not lose strength, because each newly excited region supports the passing wave to the full extent of its resources. As in other excitable media, a single pulse (e.g. the first one emitted from each pacemaker in Figure 7.5) also can propagate in the malonic acid analogue as reliably as does a periodic wave train.

Chemical Rotors

It was in the year of Belousov's death, 1970, that these crisp, colorful waves were first described clearly. Later that year I wrote to Zhabotinsky inquiring whether such waves were always concentric closed rings, as I had a mathematical interpretation that required them to be. In particular, I wondered, did he ever see all the wavefronts linked into a single continuous spiral whose inner endpoint rotated about a phase singularity? I thought my mathematics forbade this. This image came to mind because biochemical clocks produced such patterns in fungi in my laboratory [Winfree 1970b; Winfree 1973a], but in a simple chemical medium it seemed unlikely. Scientific correspondence with Russia is often slow, and I lost patience. Securing a bottle of ferroin, I mixed up Zaikin and Zhabotinsky's recipe and beheld spirals everywhere! (Figures 7.6, 7.7) Not

Box 7.A: The Malonic Acid Reagent: An Aqueous Solution of the Nerve Equations

By an apparently minor adjustment of Zhabotinsky's recipe, the reaction can be made to quit oscillating spontaneously. But its "excitability" and capacity to propagate chemical signals remain uncompromised. This discovery [Winfree 1972a] came as a surprise to interested mathematicians at the time because understanding of these chemical waves was based instead on the new reaction's other novel property: that it is a liquid clock.

However, this was no surprise to physiologists, accustomed to the same behavior in living excitable media: the delicate readiness of neurons to fire in response to a small stimulus can usually be adjusted until the "stimulus" is too small to notice. Then firing occurs spontaneously, and again as soon as the refractory interval is over, and again, and again.

The analogy goes deeper, to the equations governing this so-called malonic acid reaction and to the Hodgkin-Huxley equations that govern the electrical membranes of living organisms [Winfree 1974c; Troy 1978]. They are so close in form and in qualitative behavior that the malonic acid reagent might be fairly called "aqueous solution of the equations of electrophysiology" (pun intended). As such, it provides a convenient tool for teaching and even for research into the implications of excitability in two-dimensional and three-dimensional contexts. Figure 7.5 shows one such implication: propagating circular waves. Another, as we saw in earlier chapters, is that the right stimulus can start a perniciously persistent wave that has no source but just circulates around a phase singularity.

This blue spiral wave (lighter grey in these black-and-white halftones) is spun out like water from a rotating lawn sprinkler. The sprinkler head is a tiny configuration of chemical gradients—the rotor of Chapter 5—that rotates itself while stimulating adjacent liquid on one side to turn blue. A spinning rotor excites the surrounding liquid, so waves—or more exactly, a single wave—propagates outward as a spiral (Figures 7.6 and 7.7).

Photographs like these are hard to make, since the slightest disturbance of the reacting liquid smears or distorts the pattern developing in it. But the reagent can be adsorbed into the submicroscopic pores of chemically inert membranes, where it is mechanically bound by its own viscosity, and wave patterns are accordingly immune to handling. Such membranes can then be dried, to be wetted again only when waves are to be brought back to life. In this convenient form (recipe in Winfree 1980), the malonic acid reagent lends itself to many striking demonstrations of the peculiar properties of excitable media. Waves can be started in concentric rings by repeatedly "throwing a stone in the pond" by a touch of a silver toothpick. By laying down a temporary barrier of ions with a paintbrush, a segment of wavefront can be erased, leaving two broken ends of blue wave that become counter-rotating phase singularities. To understand why "erasure" is possible and why endpoints remain where created, more detailed analysis of the underlying kinetics and chemical gradients is required [Winfree 1978]. Rotation can be slowed down, and wavelength increased by laying the paper on a cold surface. Exposing it to a hot lamp has the opposite effect. Ultraviolet light serves as a stimulus. By tilting the paper under a lamp, a graded stimulus is administered, generally transverse to existing waves. The outcome for a localized stimulus far from the boundaries of a uniform medium, as foreseen in Chapter 6, is typically a pair of mirror-image rotors (Figure 7.8).

Figure 7.8: On a 1-cm square of millipore filter soaked in malonic acid reagent, a pair of mirror-image rotors have been created by a gradient of ultraviolet excitation falling across a gradient of timing as in Figure 6.1 (adapted from Winfree 1985, with permission). As mentioned in Chapter 6, a stimulus that does not overlap the edges of a uniform excitable medium devoid of singularities can create new ones only in opposite pairs.

Figure 7.9: Spiral waves of cAMP release are optically revealed by cell-shape changes in a dense layer of about 50 million social amoebae piled about 2 cells deep. The dish is 50 mm in diameter, roughly as in the previous two figures. Wave spacing is about 2½ mm. There are about 130 cells, 6 microns apart (each 10 microns in diameter), across the diameter of the rotor at the source of each spiral. The rotor thus contains about 13,000 cells. Reprinted from Newell 1983 by courtesy of Marcel Dekker, Inc.

reading the Russian literature at the time, I was unaware that this is exactly what Valentin Krinsky, in the Puschino Biophysics Institute of the Soviet Academy of Sciences, had anticipated years earlier in any excitable medium: reentrant excitations analogous to those imagined in heart muscle [Krinsky 1968]. Needless to say, Zhabotinsky too had seen them: they appear as a black-and-white photo for the first time in his thesis [Zhabotinsky 1970, in Russian], for the first time in color in a Swedish popular science article [Winfree 1971], and for the first time in English in *Science* in 1972 [Winfree 1972a].[4]

Thus it was discovered that the malonic acid reagent also supports rotors,[5] as we first saw hinted in the "pinwheel" phase-resetting experiments and then saw in heart muscle. For more detail, see Box 7.A.

As with most seeming novelties, it turns out that rotors were seen in a different context just as clearly long before. They are in fact a common,

[4] The Puschino laboratory has prepared a splendid 35mm movie film in celebration of their discovery and of its analogues in electrophysiology. The USSR Academy of Sciences was kind enough to convert it to English and 16mm for me: it is available for loan.

[5] It is necessary to standardize vocabulary. I will use "rotor" to mean the tiny rotating source at the "core" of a "spiral wave," and use "pacemaker" to mean the variably longer-period sources of concentric ring waves. As noted in Winfree [1985], the same words have been used differently elsewhere and different words ("reverberator," "leading center," "leading circle") have also been used, not always consistently, for combinations of the same ideas.

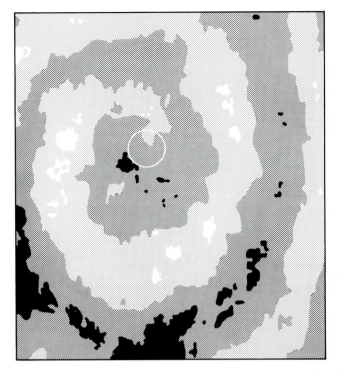

Figure 7.10: The biochemical structure of Figure 7.9 is revealed by direct assay for cAMP on a 25-mm² field containing 120,000 amoebae about 14 microns apart on average. (Each dot of the stipple used here represents about 2 amoebae.) Concentration ranges from 0 to 1 micromolar, indicated roughly by brightness of stipple. Note the irregular local maxima and minima along the spiral crest lines as foreseen in the generic Figure 2 of Winfree 1972a. Wave spacing here is about 1.7 mm; the rotor, with that perimeter, is 0.5 mm in diameter and so contains about 1000 amoebae, 36 across the diameter. Its guessed position is indicated by a superimposed white ring. Wave velocity leaving the rotor is about 0.3 mm/min (5 micron/sec). Adapted from Tomchik and Devreotes 1981. The original figure used a 9-color code for concentration (here reduced to 4 levels of shading), not to be confused with the use of continuously graded color throughout this book to represent phase in a cycle.

perhaps universal, phenomenon in excitable media, whether the medium is capable of uniform spontaneous oscillation or not. For example, would you believe that Figure 7.9 is not the malonic acid reagent?

Rotating Biochemical Waves in Living Media

The spirals in Figure 7.9 are rotating[6] at about the same period as the more familiar chemical wave, with about the same speed and spacing, but the medium they inhabit is alive: this is a lawn of living cells of a

---

[6] Strictly speaking, the spiral wave surrounding a rotor doesn't "rotate," but propagates forward everywhere, all along its curving wavefront. Only because of the particular shape

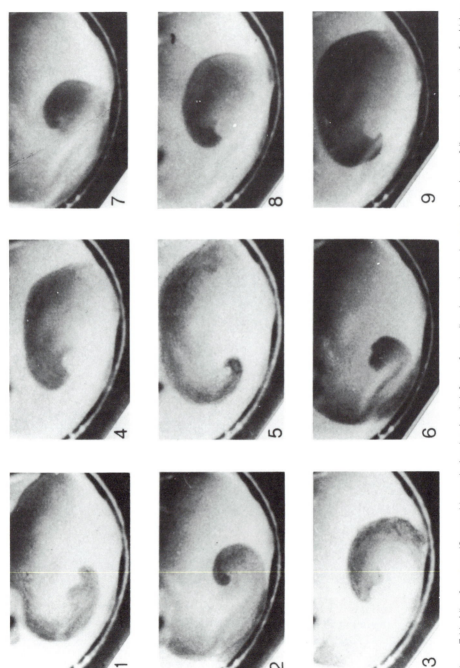

Figure 7.11: Nine frames at 40-second intervals showing the dark front of spreading depression as it rotates through two full turns on the retina of a chicken. (Adapted from Gorelova and Bures 1983.) The retina is 15 mm in diameter; the field of view here is about 1 cm².

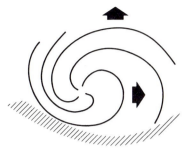

Figure 7.12: Contour lines sketched on Figure 7.11 indicate isochrons (wavefront positions) at intervals of 1/4 cycle.

slime mold [Bonner 1967]. These individual cells, called social amoebae, amount to little droplets of excitable gel that communicate through the film of liquid on their culture medium. We witness here, perhaps, a living fossil replaying events that were common during evolution from unicellular to multicellular organisms two billion years ago.

These cells will not cooperate to propagate waves when they first hatch from individual spores. But at a certain stage in their life cycle (viz., when they run out of bacteria to eat), they desist from their former solitary wanderings and become gregarious. Each begins to synthesize and hoard molecules of a hormone called cAMP, to be released later in an abrupt "sneeze." Neighbors sense it and sneeze in turn, adding still more to the cloud of cAMP diffusing through the communal liquid film (Figure 7.10). The contagion propagates like a fire on a lawn of dry grass, leaving each cell depleted and refractory but beginning again to accumulate a new hoard of cAMP.

What is seen in Figure 7.9 is not actually the wave of sneezing, but a wave that follows just behind the sneeze: cells deform momentarily, changing the way they refract light, right before they take a step toward the source that triggered these changes. This step into the advancing wave, repeated wave after wave, brings the cells together near the organizing center, the singularity. There they come in contact with one another. They assemble themselves into a slug-like organism and crawl off to some suitable place to complete their life cycle by erecting a little "fruiting body": some cells sacrifice themselves to become its dead stem, holding aloft the elect, who become spores and scatter to hatch again as solitary amoebae.

This process has been familiar to biologists for many decades, but its

of the wavefront, inherited from the rotating source, does this propagation resemble a rigid rotation. The difference is apparent where some local distortion of the spiral wavefront is seen to propagate along local normals, rather than being carried sideways as though engraved on a phonograph record.

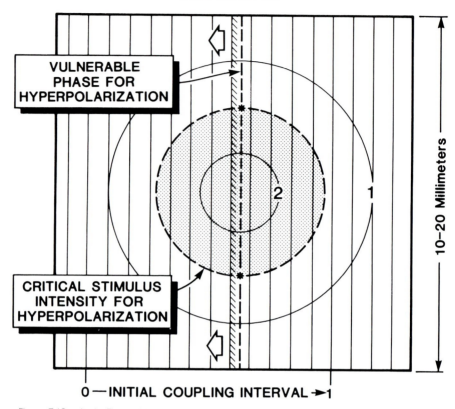

Figure 7.13a: As in Figure 6.1, but this time an inhibitory (in electrical terms, hyperpolarizing, anodal) stimulus is given. For convenience its center is placed among pre-stimulus isochrons near the action potential's upstroke, where the hyperpolarizing singularity or phase discontinuity is commonly found in experimental preparations (Chapter 4).

frequent organization by rotating spiral waves was documented for the first time only in 1965, in the laboratory of Gunther Gerisch in Germany [Gerisch 1965].

Why spirals? As in malonic acid reagent, *Renilla*, heart muscle, and other cases to follow, concentric ring waves also occur. As in those other cases, the concentric ring pacemakers have quite diverse periods, but the spirals rotate at a common period typical of the medium. Probably the ring waves come from irregular spots: ectopic foci in the heart, dust nuclei in the chemical reagent, and spontaneously active minority cells among the amoebae. The spirals probably come from rotors constituted of a near-continuum of cells whose reactions are dynamically linked through cAMP diffusion and membrane receptors.

Like heart muscle and chemical oscillators, social amoebae can also be induced to oscillate spontaneously; their phase-resetting curves can

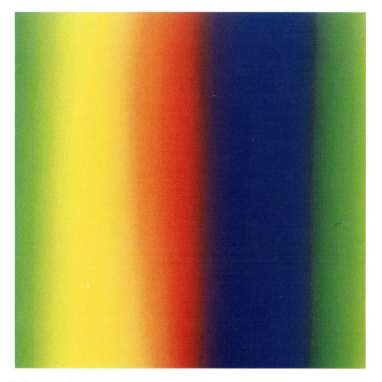

Figure 7.13b: As in Figure 6.1b, the pre-stimulus isochron field is reinterpreted by our standard color code, red representing the action potential and purple-blue its wake.

then be measured. Even-type resetting is observed in response to realistic stimuli both in the malonic acid oscillator (see above) and in these bio-chemical oscillators [Malchow et al. 1978]. This implies that a suitably graded stimulus of the same kind would be sufficient to initiate rotation around twin singularities in a lawn of such cells, as in prior pinwheel experiments. If that is how the slime mold's rotors get started, no one knows just where the stimulus might come from. It is more likely in this case that they arise from waves broken not by functional holes of tem-porary inhibition or refractoriness but (as in tachycardias turning to reentry, perhaps) by rupture of wavefronts arriving from distant sources at intervals shorter than the refractory period of local cells.

Rotating waves are also directly visible in the retina of the eye, a network of nerve cells that many regard as an outcropping of the brain surface. Figure 7.11 shows two full turns of a spiral wave in a chicken's retina, photographed at intervals of 40 seconds in the laboratory of Jan Bures in Prague [Gorelova and Bures 1983]. Figure 7.12 shows isochrons

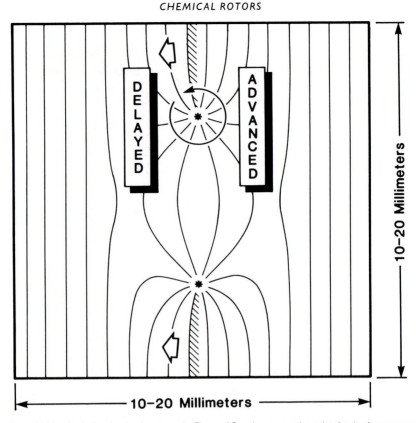

Figure 7.14a: Post-stimulus isochrons as in Figure 6.2, using a negative stimulus in the same numerical caricature (see Appendix). In this arrangement the singularities are formed by erasing an arc of latency zero (isochron 1.0, the wavefront) and less conspicuously creating a new arc of every isochron in a range surrounding 0.5. (The positive-stimulus version of Figures 6.1 and 6.2 inconspicuously erased arcs of pre-stimulus isochrons near 0.5 and created a new arc of wavefront isochron 1.0 and nearby.)

sketched along successive wavefronts, converging to the central singularity. Unlike cardiac action potentials (typical speed 1 meter per second), waves from this chemical rotor propagate at only 3 mm/min. This sequence was filmed during studies of spreading depression, a slow wave of massive depolarization typical of sheets of neural tissue such as the retina and the cortex of the brain. When extracellular potassium concentration exceeds about 12 millimolar, the electrochemical polarization of cell membranes is so diminished that the fast sodium channels open completely: sodium ions enter, forcing out equally many potassium ions, thus worsening the situation until saturation is reached. The condition spreads like an epidemic by diffusion of extracellular potassium. Metabolic pumps restore equilibrium minutes after the wave passes. (These

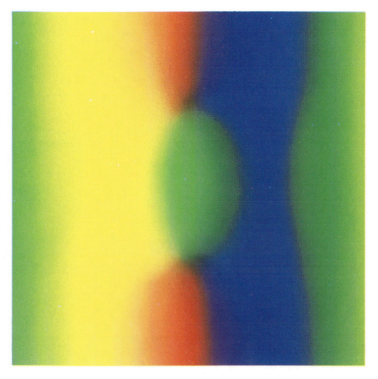

Figure 7.14b: As in Figure 6.2, the post-stimulus isochromes are shown as a continuous color field. Red is action potential, purple-blue its wake.

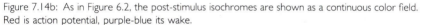

waves, incidentally, are involved in migraine headaches [Lauritzen et al. 1983]; they were first observed by the pioneering physiologist K. S. Lashley [1941] as moving boundaries of a blind zone in his field of vision during migraine attacks.)

How are such vortices initiated? Consider first the analogy of stones thrown in a pond (Box 7.A). By throwing two stones, waves are sent forth to collide midway between and (considering that this is a "pond" not of water but of an excitable medium) to erase one another. A stone thrown just at the edge of the refractory wake behind a previous wave creates a semicircle of excitation, its two endpoints becoming counter-rotating stationary singularities. Diagrammatically, this resembles Figures 6.1 and 6.2, reinterpreted as follows: the beach around the island of strong (even) resetting is the outer edge of effective excitation; the excitatory critical phase (the depolarizing, cathodal vulnerable phase) is the end of refractoriness and beginning of reexcitability. In computations on

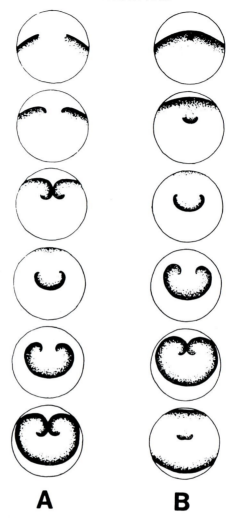

A   **B**

Figure 7.15: In column A, in six successive sketches (from photographs of a 15-mm diameter disk at 30-second intervals), the development of a mirror-pair of phase singularities is shown after anodal stimulation, presumably near the hyperpolarizing vulnerable phase, just ahead of an arc of spreading depression wavefront in chicken retina; column B differs from A in that the stimulating electrode was a cathode centered behind the wavefront near the presumed depolarizing vulnerable phase (near the end of repolarization). Adapted from Agladze et al. 1983 and Gorelova and Bures 1983.

Box 7.B: Meander

The first movie films of spiral waves in the malonic acid reagent [Winfree 1970d] showed breathtakingly regular spacing and timing of the wavefronts spinning out from many rotors. But careful examination of the inner endpoints of those spiral waves revealed that they meander irregularly [Winfree 1973bc] (Figure 7.16). This was thought to be an artifact of irregular heating or imperfectly stirred initial concentrations. It wasn't. Successive improvements of experimental technique rendered no improvement in the rigidity of core rotation. These observations seemed all the more bewildering when the first computations of persistently rotating excitation in an excitable medium described by differential equations [Winfree 1974ab] produced an absolutely rigidly pivoting, very smooth gradient of concentrations. Many others have since computed rotors, also obtaining rigid rotation [references in Winfree 1985], and have filmed malonic acid rotors to describe meander in more detail [Agladze et al. 1983].

But the mystery can also be described in reverse: some computations of diverse excitable media have now produced irregular meander [Kogan et al. 1980; Rossler and Kahlert 1979; van Capelle and Durrer 1980—see Figure 5.3; Miura and Plant 1981; Pertsov et al. 1984; Zykov 1984]. And Müller, Plesser, and Hess [1985] have demonstrated absolutely rigid pivoting of the rotor in malonic acid reagent (Figure 7.17).

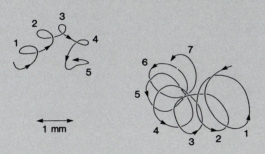

Figure 7.16. Left: A rotor in the malonic acid reagent [4 February 1973, recipe of Winfree 1972a] as free liquid but with half the usual acidity to diminish excitability. The inner edge of the spiral, close to the nominal pivot point, circulates in roughly 120-second loops. Speed varies along the loop. Right: One of seven rotors (24 December 1976) showing meander along roughly 30-second loops in millipore filter material saturated with same reagent (standard acidity) under oil, 24° C. Three of the seven precessed opposite to looping, others indistinct in this regard. The distinct ones follow Agladze's Rule, as do all that close neatly in multiloop cycles, e.g. trefoils.

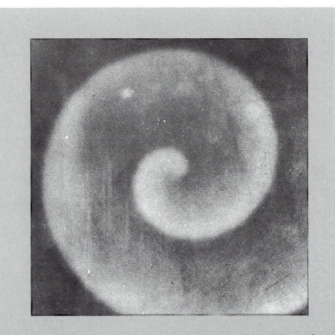

Figure 7.17. A single snapshot of the rotor and surrounding spiral wavefront in Belousov-Zhabotinsky reagent in 450-nm light. Optical density indicates the ratio of reduced to oxidized ferroin. The square is 6 mm on edge; wave spacing is 2 mm. Adapted from Winfree 1973c, Figure 3.

According to Zykov [1984], there is no mystery: rotors may pivot or meander according to details of the local kinetics. As noted in Chapter 6, the first observed rotors in atrial muscle were strictly periodic; those seen since (under somewhat different physiological conditions) meander. Meander occurs if the initial rise of excitation is too quick and/or pulse duration is too long, in the models studied numerically by Zykov.

Are there regularities about the path followed in meander? Gorelova and Bures [1983] describe the path of a "spreading depression" rotor in chicken retina as a "cycloid." Agladze et al. [1983] report that the inner end of the spiral follows clockwise loops that don't quite close, while the loops precess around a larger anti-clockwise circuit (or the mirror image of both), as in Figure 7.16 from my lab books. However, many entries therein are less distinct, the loops overlapping heavily and irregularly; others close perfectly in three loops. These and a rich variety of other paths result from computations using diverse excitable media [Nandapurkar and Winfree 1986].

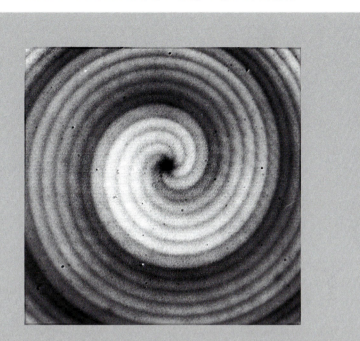

Fig. 7.17, cont. Six superimposed snapshots like above in 490-nm light at 3-second intervals during 17.4-second rotation of a spiral from Müller, Plesser, and Hess [1985]. (The six appear to differ in darkness only due to an artifact of the numerical processing.) The square is 4.6 mm on edge; wavelength is about 1.2 mm, so nominal core radius is $1200/2\pi = 190\mu$. A black region of about half that radius surrounding the phase singularity shows relatively little change of optical density. Molecular diffusion across the core radius takes about $(0.019 \text{ cm})^2/2 \times 10^{-5} \text{ cm}^2 \text{ sec}^{-1} = 18$ seconds. Reprinted by permission from the authors and the AAAS. Copyright 1985 by the AAAS.

comparable excitable media, this method was used by Kogan et al. 1980, Figure 3, and by van Capelle and Durrer 1980 (see Figure 5.3, above).

Alternatively, also as in Box 7.A, a temporary barrier to propagation may be placed just ahead of a wavefront to open a gap in it. The two edges of the gap become counter-rotating singularities. Diagrammatically (Figures 7.13 and 7.14) this resembles Figures 6.1 and 6.2, but reinterpreted as follows: the beach around the island of strong (even) resetting in Chapter 6 is here the boundary of the region inside which phase zero (firing) cannot occur; the inhibitory critical phase (the hyperpolarizing, anodal vulnerable phase) in Chapter 6 is here the moment just before firing when it is already too late to inhibit. In computations on similar

Box 7.C: What Is "Excitability"?

The vague idea of "excitability" has pervaded our discussion in this book, just as it has pervaded the scholarly literature for two decades. The idea seems clear: a reaction is excitable if it has a unique steady state that the system will approach from all initial conditions, but there exists a locus of initial conditions near which either of two quite different paths may be taken toward the unique steady state. If one of these paths is a lot longer than the other, then the system is "excitable." But no one has yet offered an exact definition: only examples. As more and more examples have come under study it has become apparent that there are different degrees and kinds of "excitability." Marginal cases—media in which propagation of the front is barely stable or in which recovery to the initial quiescence is extremely slow along one path, for example—support qualitatively different wave phenomena. Whatever the mathematical criteria may be for stable rotors, singular filaments, and organizing centers, they will not be discovered by attention to any single incarnation of the idea of an excitable medium. They might be discovered by attention to extreme cases, the limits beyond which local kinetics cannot be distorted without loss of these conspicuous features.

excitable media, this method was used by Gul'ko and Petrov 1972 and Shcherbunov et al. 1973.

These verbal translations from phase-language to excitability-language are less than compelling. The correct language is the geometry of multi-dimensional state space. The equivalence of Figures 6.2 and 7.14 may be more evident in such terms (see Appendix).

Figure 7.15 shows both procedures implemented in experiments on spreading depression in the chicken retina [Gorelova and Bures 1983, Figure 2; Agladze et al. 1983, Figure 20]. In column B, a cathodal stimulus just behind a passing front excites a semicircular wave whose two endpoints become mirror-image rotors. In column A, an anodal stimulus inhibits just before the wave arrives, creating a gap whose edges, once again, become mirror-image rotors.[7] Either way, the effect might be described still more simply: Given a wave, an appropriately graded stimulus may erase part of it, leaving two dangling endpoints of wavefront. As it happens, such endpoints tend to quit propagating, remaining rooted

[7] I do not understand why this current, applied extracellularly, does not excite spreading depression like the cathodal current, since both currents must pass through cellular membranes in both directions.

where created while distant parts of the wave continue as though nothing had happened. The result is a pair of spirals.

Returning to Figure 7.11, in the course of many further revolutions, the rotor slowly drifted across the bird's eye, possibly transverse to a of excitability (see Box 7.C). These modes are typically periodic except at points *where time breaks down*: at phase singularities where a full cycle of isochrons converges to a pivot. The pivot is a wave source, organizing periodic activity in time and in space throughout the medium.

As we saw in the previous chapter, such objects are very hard to study directly in heart muscle. Muscle tissue is full of irregularities that further complicate patterns that are already complicated enough in themselves. large-scale gradient in the electrical properties of its cells. Such migration is also typical in heart tissue and in computational caricatures of chemically excitable media exposed to a gradient of illumination, acidity, or temperature. In fact even in perfectly uniform excitable media, rotors can migrate spontaneously in a random way (see Box 7.B) called "meander," even when undirected by gradients [references in Winfree 1985; especially Kogan et al. 1980; Zykov 1984]. Meander is especially conspicuous in cardiac muscle.

These properties and others are typical of a wide variety of media Moreover, such waves recur several times per second, traveling at speeds called "excitable," any one of which might be regarded as an analogue computer for exploring the modes of excitation of idealized equations on the order of a half-meter per second. And they are invisible. To detect them electrically requires implanting an array of electrodes backed up by a substantial computer (a procedure that has been used also in the chemical reagent, incidentally; see Botre et al. 1983). In contrast, the chemical waves are slow, colorful (the pale yellow cerium being replaced by ferroin, a red/blue indicator dye), and plainly visible in three dimensions. Moreover the chemical medium can be made as "lumpy" and discontinuous or as ideally uniform as desired; the latter has obvious advantages in the first stages of dawning awareness.

We turn next to exploit the most conspicuous advantage of this chemically excitable medium: its transparent three-dimensionality. Having likened chemical rotors to tornadoes and to cardiac reentry, it now behooves us to note that neither chemical solutions nor tornadoes nor hearts are two-dimensional (Figure 7.18).

Figure 7.18: Vortices in three-dimensional media are not mere points, but one-dimensional filaments. In the case shown (giant cyclonic waterspout, observed during Lower Keys Waterspout Project, 1969, by aircraft 14 naut. mi. north of Key West, 9/10/69 by J. H. Golden, NOAA/National Weather Service), the filament vanishes at a lower interface and presumably becomes diffuse in the thinner upper regions of the medium. In a uniform chemical reagent subject to only local disturbances far from the walls, those evasions are not convenient. What becomes of the ends of the filament?

PART III

Organizing Centers

CHAPTER 8

Patterns of Timing in

Three-Dimensional Space

We ever long for visions of beauty.
Ever we dream of unknown worlds.
—Maxim Gorki

This book began as an exploration of biological clocks, their resetting, and their arrhythmias. The world being made of both space and time, it proved impossible to evade embroilment in spatial patterns of timing. The example most emphasized was taken from electrophysiology: wave patterns in the heart and their arrhythmias (Chapters 2–6). We are thus inescapably embarked upon an inspection of periodic patterns of timing in three-dimensional space. This is not an easy task. For expediency we seek first to recognize the patterns typical of any uniform excitable medium, forgetting about the idiosyncrasies of that particularly complicated excitable medium. Accordingly we redirected our attention in Chapter 7 to relatively uniform biochemical and neurochemical media, and to a perfectly uniform chemical soup. What kinds of structure most naturally arise and maintain themselves in such excitable media? How can this query best be posed?

The first step, which no theorist would have anticipated, is to set aside the mathematical literature of reaction/diffusion pattern formation. It is ironic that the first convenient and understandable laboratory example of reaction/diffusion structures violates the central mathematical principle pursued by developmental biologists [Meinhardt 1982] and physical chemists [Nicolis and Prigogine 1977] since Alan Turing's seminal paper of 1952 [Turing 1952]. That principle describes a patterned instability of the spatially uniform reaction, which owes its rate of growth to inequalities among diffusion coefficients and its final pattern largely to boundary conditions. It has no unique source, but consists of a periodically structured field, every part of which is equally important. In contrast the Belousov-Zhabotinsky reagent (as modified [Winfree 1972a]) is perfectly stable in its uniform quiescence. The pertinent diffusion coefficients are equal within a factor of two; the role of diffusion is chiefly to enforce local coherence on patterns caused mainly by reaction. The characteristic reaction-diffusion patterns are localized and particle-like; boundaries further away than a small fraction of a wavelength have little influence on rotors, shielded inside a field of outgoing waves.

It is ironic also that a revolution in our understanding of biological morphogenesis seems to be coming about through abandonment of classical reaction-diffusion ideas for something closely related but distinctly, qualitatively different: a *mechano*-chemical analysis of excitability in the three-dimensional cytoskeleton of living cells [Odell and Bonner 1986; Oster and Odell 1984ab; Oster et al. 1983; Oster 1984; Oster et al. 1985]. Amerigo Vespucci was right on target when he commented in 1500 about a recent expedition to the New World, "Rationally, let it be said in a whisper, experience is certainly worth more than theory."

The two most heroic strides toward an understanding of morphogenesis in reaction-diffusion processes were taken in the same year (1951) by an Englishman and a Russian who never met. The Englishman committed suicide before his work was appreciated; the Russian was refused publication and died before his work was known outside a small circle of colleagues. The Englishman's work provided the nucleus for a ponderous industry of theoretical elaborations. The Russian's ultimately provided the clearest example of self-organization by reaction and diffusion, but entirely different concepts are needed to understand it. Those concepts, so far as they have been explored in English, are presented in two-dimensional context in Winfree 1978, Winfree 1980, in the preceding chapter, and in two recent collections of essays: Krinsky 1984 and Field and Burger 1985. The former provides especially accessible guidance to the seminal Russian literature.

The leap from two to three dimensions merely elaborates the foregoing with more geometry. In this chapter and the next we will not deal explicitly with local reaction kinetics, be they oscillatory or excitable or both, or with molecular diffusion. We will simply contemplate the geometry and topology of a volume of medium organized periodically in space and time. In appropriate cross section, those patterns are required to be the spiral waves already found to be stable in two dimensions. That's it: nothing more complicated. But the results are surprisingly rich.

Phase Singularities Again

If thou, dear reader, art wearied with this tiresome method
of computation, have pity on me, who had to go through it
seventy times at least, with an immense expenditure of time.
—Johannes Kepler,
Astronomia Nova (1609)

A topological trick served us well in coming to grips with the timing of circadian rhythms and the spatial patterns of timing in heart muscle. We used it in two versions: as a comment on the integer winding number of phase along any closed path and as a comment on color wheels and the impossibility of smoothly coloring their insides from the same palette. Both versions will serve us again now as we explore periodic arrangements of timing in space that are natural to excitable media. Except for uniform quiescence, such arrangements consist of concentric bag-like waves emerging from peculiar chemical structures called organizing centers [Winfree and Strogatz 1983abc, 1984ab]. It turns out that there are distinct species of organizing centers, each qualitatively unlike the others, like atoms in the periodic table of the chemical elements [Winfree and Strogatz 1984b]. Because only the first one predicted, the "hydrogen atom," so to speak, has been sought (and found) in laboratories, we won't venture further afield in this chapter. Chapter 9, however, will include computer graphics anticipating the next more complicated organizing center (the "helium atom").

A review of the principles of periodic organization involves, first, the idea of phase in a periodically active medium. Phase represents the fraction of a cycle elapsed since the last passage of a wavefront. We might color-code a snapshot according to phase, starting with red as before, changing through purple, violet and blue to green, yellow, orange, and red (see Figure 5.5). The medium lingers in the orange-red phase until an encroaching wavefront triggers it to execute another cycle.

While listing these abstractions, I am really thinking of the malonic acid reagent (e.g. Figure 7.6), and perhaps by now so are you. That is why I assigned orange-red to the quiescent phase ($0 = 1$) and violet to early excitation. But care must be taken to avoid further confusing the phase-label colors with the actual colors of the chemical soup. Putting different indicator dyes into the brew will change the color cycle without changing the timing relationships that we encode by the color-wheel sequence. If there is only a single visible indicator (the usual recipe), then there is not even a full cycle of actual chemical coloration: color, being a single chemical concentration, just changes and reverts back through the same sequence in each period. Nonetheless, one period still represents a full cycle of phase, encoded here by a complete color wheel of labeling hues.[1] Figure 7.6 shows a pair of spiral waves, each emerging from a central rotor. Figure 8.1 is a black-and-white photograph of a left-handed rotor and the spiral wave departing from it, color-labeled according to our code (not according to the indicator dye) to display phase. The wavefront in this inner region is nearly an involute spiral, the only mathematical curve compatible with:

(1) local wave propagation perpendicular to the local wavefront at uniform speed (outside the wave source, the central rotor), and
(2) uniform rigid rotation of the whole pattern in pace with the steady gyration of its source, the central rotor.

This is very nearly the shape of a coiled rope. More exactly, it is the curve traced by a pencil tethered to a cylindrical post: as the string winds up about the post, the pencil traces a spiral with spacing equal to the post's circumference. In the very confined spaces available to rotors in nerve or muscle tissue there is scarcely room for one turn of the spiral: just a curvy fragment is seen. It does not particularly resemble an involute because the confining boundaries interfere with propagation. But even in the chemical reagent with boundaries far away, the involute is an inadequate caricature to the extent that assumption (1) fails: the wave does not propagate with normal speed where it is (necessarily) sharply curved near the pivot. Propagation speed is less where the wavefront curves more sharply [DeSimone et al. 1973; Zykov and Morozova 1979; Zykov 1980; Zykov 1984]. As a result the wavefront curves inward near the center more abruptly than does an involute (e.g. see Winfree 1980, Figure 13-3). A definitive mathematical analysis has not yet been constructed, but tantalizing approximations are converging on the solution [Koga 1982; Hagan 1982; Mikhailov and Krinsky 1983; Krinsky and Malomed 1983; Kuramoto 1984; Fife 1984; Welsh and Gomatam 1984; Malomed and Rudenko 1986; Keener and Tyson 1986]. The involute approximation is also inadequate to the extent that assumption (2) fails: depending on the recipe (or on the parameters of a numerical simulation), the rotor may simply rotate as in (2) or it may also translate as it rotates, moving along a loopy path, possibly closed. Not even an attempt has yet been made to understand this analytically.

---

[1] A single dye makes only a single component of the chemical cycle visible to the human eye; as that dye changes and changes back, we see each color twice in each cycle. If the concentration of another colored substance is meanwhile waxing and waning out of phase with the first, then each moment in the cycle becomes uniquely colored. This is the case in principle, but the "other substances'" absorption spectra are mostly outside the range of human visual sensitivity.

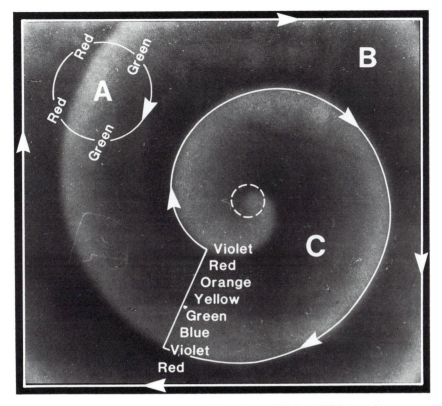

Figure 8.1: A rectangle of malonic acid reagent (for recipe see Winfree 1972a) about 4-mm square contains a rotating chemical structure about 1/3 mm in diameter (outlined by dashed circle), from which a left-handed spiral wavefront radiates. Except within that rotor, every area element progresses through the same cycle of oxidation and reduction. Phase in the cycle could be color-coded everywhere as in Figure 5.5; the colors are here indicated only in writing. The winding number is +1 along clockwise closed path C and along path B. The winding number is 0 along path A.

Why do I draw attention to the fact that even this fifteen-year-old problem still evades quantitative solution? Only to underscore the urgency of finding some more expedient style of exact analysis. We have in fact already found and used it throughout the preceding chapters: topological analysis of timing relationships. We begin by defining "phase" in Figure 8.1. The photograph should be imagined to be overlaid by colors encoding the phase at each point; thus every spiral parallel to the wavefront would be uniformly colored. The sequence of colors is indicated just along one sample radius, from one wavefront to the next. Curve C is a closed path encircling the pivot. We now conduct a clockwise tour of inspection to check the phase (hue) at each point. It will be easiest to imagine the tour conducted very quickly, so all phases remain as shown in this snapshot (though the result is the same if we take our time and measure local phase relative to a portable clock). Start at the red inner corner and proceed clockwise to the red outer corner, then radially back to the start. How many full cycles of color go by? Count forward changes (red - violet - blue - green - yellow - orange - red) and subtract reverse changes. The result must be an integer number of cycles, because we started on red and we end up on red. On path C the answer is +1: a full cycle has vanished, just as a day vanishes in circumnavigating the globe. On the more arbitrarily constructed alternative path B (the border of the box) the winding number is still 1. But on path A it is 0, as it would also be on any shrunken version of A, and in particular on a path so small that the color does not sensibly change along its perimeter.

What else do you notice about these three paths? The winding number simply detects encirclements of an isolated singularity: it is 0 if no singularity lies inside, and if one does, the number is ±1, depending on the direction of encirclement relative to the handedness of the singularity. The singularity is detected even though the path of local measurements goes nowhere near it! (The same argument, with citation of pertinent mathematics, was deployed in regard to putative biological phase singularities by Glass 1977. Box 10.A discusses those particular singularities. It is reviewed in mathematical terms in Strogatz 1984. For another intriguing example in a slightly different biological context, see Elsdale and Wasoff 1976 and Penrose 1979.)

This trick for remote detection of singularities must be used with care. If there are several singularities, the winding number reveals only the algebraic sum of their (+1, -1) topological charges. In Figure 8.2, path A with clockwise winding number -2 contains two more right-handed rotors than left-handed ones (3 right, 1 left). Path B with winding number 0 encloses a complementary pair, enveloped in closed-ring wavefronts. If the winding number is not zero, we detect that many unpaired singularities inside; but we never know how many additional pairs might be present until an alternative path is chosen that happens to separate them. In the pictures to follow, paths will be chosen that are small enough to contain only one singularity.

The color wheel necessarily collapses to a hueless jumble at the pivot of a spiral: thus any spiral wave, and the central rotor from which it emerges, contains an isolated phase singularity in one of two possible mirror-image orientations (clockwise or anti-clockwise). These objects are alarmingly stable, once created: they can be moved, but they cannot be destroyed by any process that leaves the winding number non-zero along a remote encircling path. Thus a rotor can vanish only by fusing with a counter-rotor, by moving to the boundary (if any) of the medium, or by succumbing to an homogenizing stimulus that engulfs a corridor

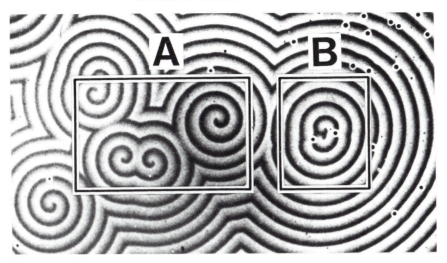

Figure 8.2: Similar to Figure 8.1, but the field of view is larger and all singular filaments are seen in projection along their length as though in two-dimensional cross section. Several rotors turn at the common 60-second period, emitting scroll waves spaced about 3.3 mm apart (at 6° C). The winding number of phase clockwise along path A is −2. Along path B it is 0. The small circles are carbon dioxide bubbles.

leading to the boundary or to a counter-rotor, thus changing the winding number to 0 on every possible path. In reverse, a rotor can be created only as one of a pair, like electron and positron, or by a stimulus that rearranges a corridor of medium that touches the boundary.

Rotors in Three Dimensions

These topological abstractions are borne out in practice in experiments with two-dimensional films of malonic acid reagent. A film qualifies as "two-dimensional" for such purposes if it is thinner than one rotor diameter. That is a quantity that cannot be known a priori, but it can be determined experimentally [Winfree 1973bc]; it is commonly about one-third of the spacing between turns of the spiral. In the rest of this chapter we turn our attention to excitable media that are distinctly not two-dimensional.

What if the pattern fills three-dimensional space? This might seem much harder to think about. In some respects it is harder, but in other respects it turns out to be quite easy. Quite a lot can be understood about the anatomy of three-dimensional organizing centers if four basic ideas are grasped [Winfree 1984b; Winfree 1985; Winfree and Strogatz 1983abc; Winfree and Strogatz 1984ab; Winfree et al. 1985]:

(1) That phase singularities in three dimensions are one-dimensional

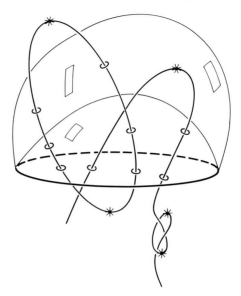

Figure 8.3: Many two-dimensional surfaces are bounded by the same horizontal ring. If the winding number is zero along that ring, then every surface must have equal numbers of clockwise and anti-clockwise phase singularities. If adjacent surfaces have similar arrangements of singularities, then the singularities must be one-dimensional filaments. Where they turn tangent to a surface, singularity pairs within the surface are annihilated/created.

filaments; they only look like points when seen in a two-dimensional slice;

(2) That they close in rings, unless the ring is interrupted by boundaries or runs off to infinity;

(3) That rings can be knotted and mutually linked; and

(4) That any number of twists (to be defined) can be locked into a ring, thus allowing topological isomers of the basic scroll ring.

Idea 1: Vortex Filaments Weaving through Space

The first of these four principles is easily demonstrated: one only need look at the malonic acid reagent in a microscope. Understanding also is easy. Imagine that Figure 8.1 is a slice through a three-dimensional volume of the reacting liquid. The surface inside path C is a flat disk bounded by C. But now that we are free to wander in three dimensions, there are many other surfaces that are less flat but have that same boundary. We might imagine a stack of nested caps made by letting the flat disk balloon out as though it were a soap film stretched across ring C. The winding-number argument applies on each of them, at every stage of ballooning: as long as the winding number of phase remains W around the border,

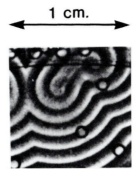

Figure 8.4: A three-dimensional scroll wave (wave spacing 1.2 mm at 25° C) around a segment of singular filament, photographed in the deeper liquid under the meniscus at the edge of a shallow dish of malonic acid reagent. From Winfree 1973b (copyright 1973 by the AAAS): this cut-out from the complete photograph is a 1-cm square.

any enclosed area must contain W colorless points (plus any number of mutually canceling pairs). Each such singularity can be followed in three dimensions as the original film billows out (and opposite pairs may appear or vanish in the process). Each singularity is thus revealed as a filament of huelessness threading its way through the colorful field of timing (Figure 8.3).

Each such hole in the color field organizes the pattern of colors around it. Being surrounded by a full cycle of colors, it is the unchanging pivot of a rotating wave that, as we saw in Figures 8.1 and 8.2, radiates away as a spiral. This thought experiment exposes the "spiral" as a scroll, seen in cross section. Figure 8.4 shows the first one ever photographed, its curvy axis several millimeters long, in the malonic acid reagent.[2]

The apparition of a scroll wave should not seem surprising. We already knew that spiral waves exist in two dimensions and that any real excitable medium has some depth, however slight. When seen in a very thin layer of reacting liquid, the scroll looks like a symmetric spiral (Figure 8.5a), but if the layer is thicker, then the scroll has room to tilt (Figure 8.5b) or to curve around (Figure 8.5c).

These scrolls consist of reacting liquid, every tiny volume element of which is alternating between different dominant reactions at the period of scroll rotation. Although timing is staggered, the cycle is otherwise the same everywhere, and therefore time-averaged rates of reaction, heat

---

[2] In 1971 this photo was halfway into the mail chute as a tiny corner of the cover photo for *Science*, illustrating an article about two-dimensional waves [Winfree 1972a]. Fortunately, a feeling that it was somehow peculiar caused its withdrawal from the chute and its replacement with the simpler photo that actually appeared. The withdrawn photo sparked an investigation of three-dimensional waves and finally appeared in *Science* the following year [Winfree 1973b].

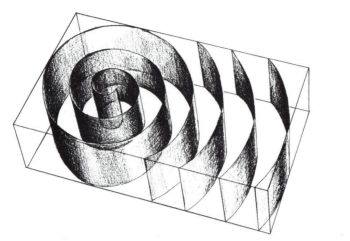

A

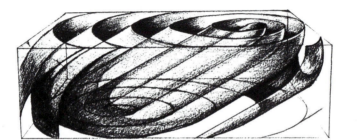

B

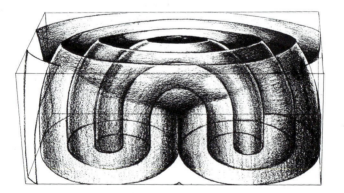

C

Figure 8.5: Scroll waves diversely arranged within a box of excitable medium (artist's conception): (a) a spiral wave projected vertically in depth; (b) as in (a) but the scroll's axis is tilted; (c) the scroll's axis is U-shaped; this is half of a scroll ring. From Winfree 1973b; copyright 1973 by the AAAS.

dissipation, and so on must be everywhere the same—except along the scroll axis. There the usual periodism is not observed. In some versions of the reagent [Muller et al. 1985], the pivot is stationary, a time-independent locus where reaction rates never get to any of the extremes attained during the cycle that continues everywhere else. There is no reason to think that time-averaged rates near this singular filament are the same as everywhere else. In other versions of the reagent [Winfree 1972a], the extremes of reaction reached periodically elsewhere are attained in the center, but irregularly, as the edge of the circulating wavefront lashes about wildly through the arrhythmic region.

The singular filament is thus a real enough object, but it is not an object like a strand of spaghetti that might be strained out from the liquid: it is only a locus of unstable equilibrium confined within the ineluctable grasp of a topological theorem (see Box 8.A).

Idea 2: Vortex Rings of Chemical Timing

Returning to Figure 8.2, the case of path B is especially revealing. Here we see two oppositely directed rotors. There might have been any number of such pairs because, pairwise, they contribute nothing to the winding number and can appear or vanish pairwise with impunity, just so they do it in a continuous way. Thus as we blow out the film to either side of ring B, we might see the mirror-image singularities draw closer together, combine, and vanish at the points of tangency indicated by sparkles in Figure 8.3. In other words, these singular filaments may typically be closed rings.

What else *could* scroll filament do? Suppose an excitable medium extends quiescently a long way in every direction, but some central volume has recently been provoked to pattern itself in scroll waves. If scroll filaments are confined to the excited volume, then they must either "peter out" like fraying ropes, or form closed rings. How can a singular filament "peter out"? If it were really a mathematical filament, it could not. Consider a sphere enclosing such a putative endpoint, and a small ring on that sphere encircling the filament's entry point. Around that ring, the winding number of phase is $+1$. The ring separates the sphere into two disk-like surfaces, meeting at their common boundary, the ring. The second surface, opposite to the first, has winding number -1 around its border, so it too must be punctured, this time in the outward direction, by a singular filament. Thus the filament did not after all end inside. But if, as in the reality of tornados, magnetic "field lines," and chemical vortices, the "filament" really has finite diameter, then this kind of argument cannot be so glibly proffered: the vortex might simply become diffuse and vanish. Does it? It has never been observed to. It seems to

Box 8.A: Is the Rotor Active, or Only Passively Driven?

An intriguing question presents itself: is the vortex core really an initiating site for the assembly of waves, and the source of waves radiating from it? Or is it better thought of as a confused region at the edge of a wavefront, a quiescent center confined near the center of the whirlwind surrounding it? Such objects do exist in other contexts. For example, fields of oscillators [Zel'dovich and Sokolov 1984] and wave fields of sound [Nye and Berry 1974; Berry 1980, 1981; Wright and Nye 1982] or of light [Baranova et al. 1983], and even the intergalactic vacuum [Kibble 1976ab; Vilenkin 1981, 1984, 1985; Winfree 1986b] all have singular filaments of topology identical to those here considered. At the singularity, amplitude is zero and phase is ambiguous, being the center of a circular phase gradient. This center is not a source; if it were, energy conservation would be violated. The zero is simply required by continuity, given the boundary conditions and the phase structure of a remote source. In hydrodynamics, too, any vortex has a core, but it is just a part of the pattern of flow, neither its cause nor a passive "effect" any more than other parts of the flow are. The eye of a hurricane or the funnel of a tornado is only part of a convection pattern driven throughout a vast area by convective instability and the Earth's rotation.

What then about a rotor in a hole-free chemically active excitable medium? Are its concentration patterns merely driven passively from the nominal perimeter along which a wave propagates tangentially?

In excitable media that are only marginally able to support wave propagation, a stable spiral wave can circulate around an expanse of quiescent medium, allegedly for the simple reason that the wavefront cannot curve sharply enough to get inside [Mikhailov and Krinsky 1983; Pertsov et al. 1984]. This "core of the core," as it is called in Russian, is virtually unstructured and in any case is too large to be integrated by diffusion during one rotation. So is there no "rotor" at the source of this outgoing spiral wave? Or is the rotor at the wavefront's edge just moving along with the attached wavefront while it rotates?

And what if the rotor is simply cut out? The outer wave still rotates, perhaps with a different period now, depending on the size of the hole [Pertsov et al. 1984]. Since its presence does markedly affect the rotation period, the "rotor" cannot be regarded entirely as a passive and dispensable appendage; on the other hand, replacing it by a hole or by inexcitable medium does leave the surrounding concentration pattern topologically unaltered, so in that sense it is dispensable or replaceable. The same is shown by inserting an impermeable barrier ("sticking a knife into its heart"): one converts the wave field to mere circulation about that obstacle, *sans* rotor. The rotor reappears, re-created by the surrounding wave field, only when the knife is withdrawn and the slit healed. Is this equivalent to saying the rotor is a passive consequence of wave geometry, rather than a wave source? I have no clear answer to this question, either.

When we escape the artificial confines of Flatland to consider three-dimensional organizing centers, the analogous question seems to have a clear answer: the organizing center is the source of waves radiating from it, cannot be replaced (without terminating wave instigation), and will not regenerate. Does this lay the question to rest? Not really. The same is true of a mirror-image pair of rotors, considered as a single entity, in two-dimensional media, yet each could be replaced by a hole without affecting the global pattern. Similarly, each ring of phase singularity in a three-dimensional organizing center could presumably be replaced by a circular thread of glass.

How can the question be answered? If there is no clear-cut experiment to resolve the matter, a suspicion arises that we have created a word game. Or is there a decisive experiment? This remains to be seen.

be a dynamical property of excitable media that vorticity tends to condense into singularities. In computer simulations, the vortex core will dilate (the rope will fray) if the reaction is suddenly made inexcitable while calculations continue with all else remaining normal; and the core will contract again to a focus if excitability is then restored [Winfree 1978]. For dynamical reasons (not topological reasons), it seems valid to infer from this topological argument with the sphere that a compact singular filament must close in a ring.

A chemically motivated thought experiment also suggests closed rings, as follows. Along the axis of scroll rotation the reaction cannot be periodic. If rotation is exact, then the axis is chemically stationary: it has constant chemical composition. Thus it lies on some (moving) two-dimensional surface of uniform concentration of each chemical involved, e.g. at concentration $A = A^*$. If only two chemicals are predominantly important, then the scroll axis lies on the intersection of two such surfaces, $A = A^*$ and $B = B^*$, both rotating on that axis. Two surfaces in three-space will intersect in closed rings (Figure 8.6) unless particular trouble is taken to arrange a more delicate intersection.

The persuasiveness of this argument rests on the supposition that only two quantities dominate, and on the observation that there is indeed a motionless axis of rotation. This observation is impressively exact in the experiments of Müller, Plesser, and Hess 1985 and in the first rotor computations at high resolution [Winfree 1974ab]. In other cases (see

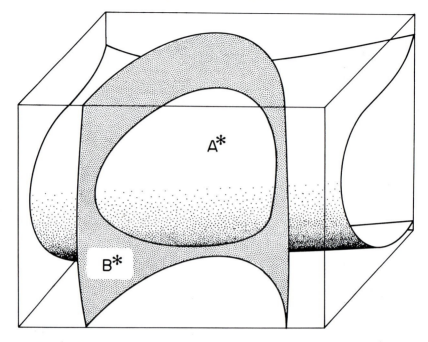

Figure 8.6: The locus of uniform concentration of substance A is a moving two-dimensional surface; similarly for substance B. The pivot of a rotor (the phase singularity) is the intersection of a particular A concentration and a particular B concentration: generically, a ring. It is a curious fact that the ring is relatively motionless while all other parts of all such surfaces propagate at standard speed. From Winfree and Strogatz 1983a.

Box 7.B) no point is chemically stationary, so one does not know what different phenomena may occur. Perhaps scroll filaments still exist and still close in rings, but must be described as perpetually gyrating and writhing; or perhaps some qualitatively different dynamical structure will provide a more apt description.

In any case, the typical scroll wave does not resemble Figure 8.5 so much as Figure 8.7, which joins two copies of 8.5c into a complete scroll ring. The singularity here is a (tilted) horizontal ring concentric to a (tilted) vertical axle. Every planar cross section that contains the axle is punctured at two points by that ring, once from behind and once from the front. Around each puncture is a planar spiral; as in Figure 8.8, they are mirror images, colliding and mutually annihilating along the axle much as in Figure 7.6. We thus come to a geometric interpretation for the pairwise creation/annihilation of rotors in two dimensions: the mirror-image rotors are only two encounters with a ring-shaped filament of singularity passing through the plane, first from one side, then returning from the other. In three-dimensional reality there are no clockwise and

Figure 8.7: Computer-drawn snapshot of a perfectly symmetric scroll ring with the foreground cut away for visibility. The inner edge of this surface lies along the rim of the rotor (a circle close to the phase singularity, slightly nicked open by the cutting plane). From Winfree 1984b.

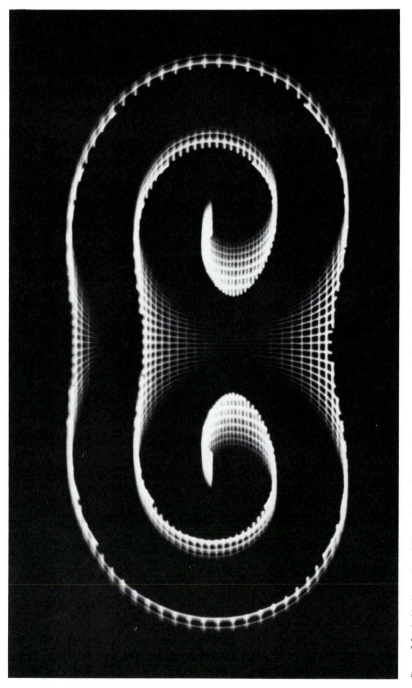

Figure 8.8: A thin-layer slice of Figure 8.7 nearly through the symmetry axis. From Winfree 1984b.

anti-clockwise rotors but only one object; the scroll ring. The scroll ring can be created and it can vanish in a solitary performance, no mirror image required.

This three-dimensional wave is a surface of revolution: Figure 7.6 spun about its midline. Cutting the scroll ring like a cheese into cross sections parallel to this axis of revolution, a series of revealing slices can be laid out in the fashion of micro-anatomists: Figure 8.9.

The anatomy of the idealized scroll ring is exposed in Figure 8.10 as another stack of serial sections, this time nearly parallel to the plane of the ring, perpendicular to the axle. Why only "nearly"? If they were cut exactly perpendicular to the axle of a surface of revolution, you would see only concentric circles: rather boring, and not representative of the less symmetric structures that might be encountered in the laboratory. Laboratory encounters are the objective. I deliberately spoiled the symmetry of these pictures in order to recognize the scroll ring in the laboratory.

Do Scroll Rings Really Exist?

What should be sought in the laboratory? In microtomed slices of a fixed and stained specimen, we look for cross sections like those above. The whole structure should also repeat itself in time at the period of two-dimensional rotors. We may expect to find some distorted version of this tidy picture in any excitable medium or tissue of clocks far from interfering boundaries. But this object is hard to demonstrate convincingly by casual inspection of a reacting volume of liquid. It has no physical substance and does not reflect light; it is only a rhythmically changing superposition of many layers of alternating red and blue water. Could it be somehow frozen and sliced up, and the serial sections stained as microscopists do with living tissues to reveal their inner complexity? In short, yes. By stimulating a point on one face of a cube of reagent, a hemispherical pulse is made to propagate toward the far wall. Beyond the wall is glass or inexcitable reagent, so the wavefront goes no further; it just continues to expand along the wall as a growing circular edge. But if the region beyond suddenly becomes excitable, then that circular edge should become a locus of rotors: a circular singular filament. This was arranged by simply pushing the wall into contact with another block of quiescent reagent. A scroll ring then developed. It is shown in Figure 8.11 as serial sections comparable to those of Figure 8.10. These are slices about 1 cm² taken 140 microns apart through a block of porous nitrocellulose saturated with the reacting liquid. After development of a scroll ring the block was chemically fixed within about one second. Notice the characteristic horseshoe sections developing into concentric inward

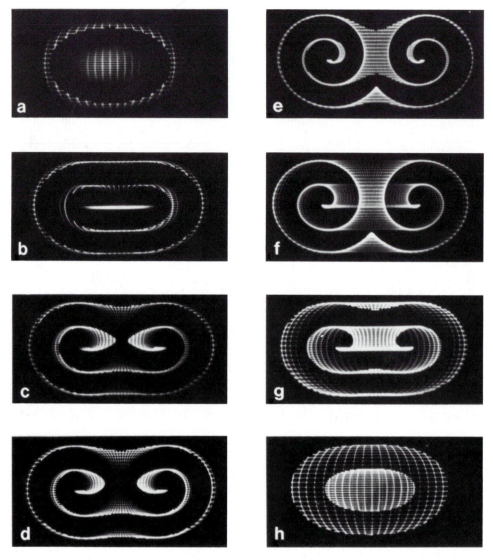

Figure 8.9: A series of parallel vertical slices through Figure 8.7, including Figure 8.8 as panel d. This scroll ring is oriented like Figure 8.14: upside down relative to all others in Chapters 8 and 9. In Figure 8.10 the same object is sliced horizontally from top to bottom, looking at it from the top. From Winfree 1984b.

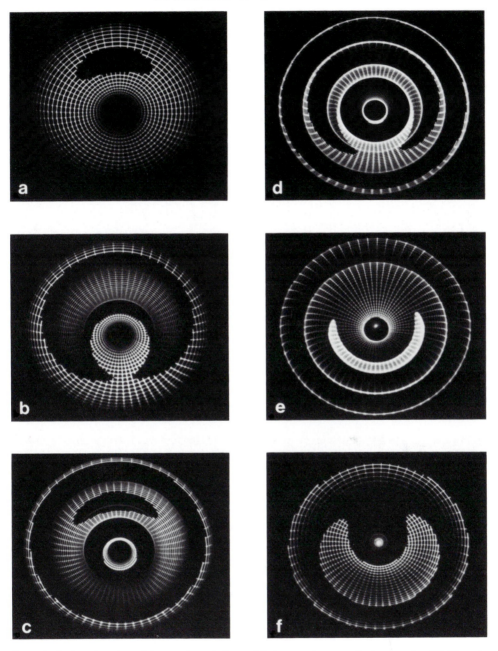

Figure 8.10: A series of parallel slices through Figure 8.7, but nearly perpendicular to series 8.9. Note the characteristic horseshoe rings. From Winfree 1984b.

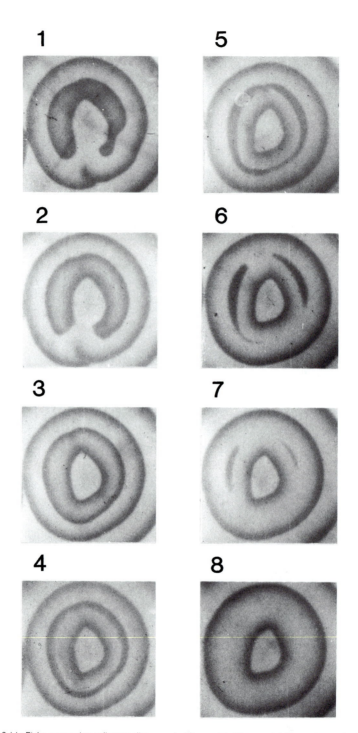

Figure 8.11: Eight successive adjacent slices, each 10 mm × 10 mm at 1/7 mm intervals, through an actual scroll ring in malonic acid reagent. Adapted from Winfree 1974d.

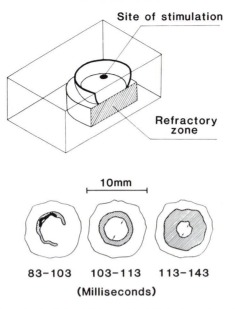

Figure 8.12: Top: The pattern of stimulation used to initiate three-dimensional reentry in rabbit atrial muscle. Bottom: Three successive intervals in the electrical activation of the atrial surface of the rabbit's heart, covering 20, 10, and 30 msec respectively. Adapted from Medvinsky et al. 1984 with permission of the senior author.

and outward rings. Copied onto sheets of nonreflecting glass and stacked up again, they reconstruct a frozen effigy of the original scroll ring. Prior to fixation, this wave repeated at the standard period of the planar spiral wave's rotation [Winfree 1974d].

Using a similar protocol, biophysicists A. B. Medvinsky and A. M. Pertsov in the Soviet Union sought and perhaps detected similar patterns of activation in a sheet of rabbit atrial muscle [Medvinsky and Pertsov 1982; Medvinsky et al. 1983, 1984]. Although atrial muscle is usually considered effectively two-dimensional, Medvinsky and his colleagues realized that it is thicker electrophysiologically than its 1–1.5 mm physical depth suggests. Because fibers lie parallel to the surface in this tissue, propagation is several times slower into the depth of the muscle than it is along the surface. In uniformly anisotropic regions of atrial or ventricular myocardium the measured factor is about 3 [Roberts et al. 1979; Spach and Dolber 1985]; in nonuniform regions, ratios as high as 9 to 12 are observed [Spach and Dolber 1985]. Thus the effective depth may be comparable to the 1-cm height and width of this preparation. This block of excitable medium (Figure 8.12, top) was stimulated to induce a reentrant tachycardia much as in the experiments of Allessie et al. [1976]: a basic rhythm was established by periodic excitation at a short

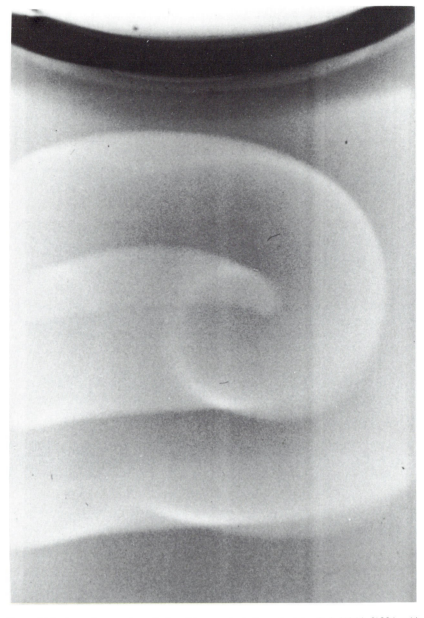

Figure 8.13: A scroll wave in malonic acid reagent photographed by B. J. Welsh [1984, with permission]. The liquid is in a 10-mm diameter test tube; the meniscus is visible at the top. The wave's edge encounters the foreground glass interface along a pair of mirror-image spirals.

interval, then a single premature impulse was applied through the same electrode during the atrial vulnerable phase (the relative refractory period after the prior basic pulse). As Medvinsky et al. conceive it in Figure 8.12 (top), the atrial muscle sheet was thus momentarily excited on top while still refractory beneath. The hemispherical region excited by the upper point-stimulus was truncated by the temporarily inexcitable lower region. When excitability recovered there, the circular edge of that hemispherical wavefront "should" have become a locus of phase singularity, the source of a scroll ring.

This is very nearly the procedure employed to make the scroll ring (Figure 8.11) in malonic acid reagent; in that chemical experiment, however, the "refractory" region below was simply glass, replaced by quiescent excitable medium to simulate recovery and thus initiate the singular ring. The outcome in heart muscle was much the same as in the chemically excitable medium. The atrium was convulsed with a tachycardia repeating at the 140-msec period typical of rotors in that medium. Horseshoe ring patterns like Figure 8.12 (bottom) were sought and found in electrical recordings at intervals of about 1 mm. Medvinsky and his colleagues suggest that they emerge to the surface of the heart from a ring-like source inside. In my opinion the present evidence remains ambiguous because of the uncertain effective depth of the preparation, the poorly documented unusual spatial resolution (1 mm), and because one cannot strictly distinguish a scroll ring from a mere doughnut-shaped wave by watching it break through a stationary window far from the source ring. A cut intersecting the ring itself would be necessary for proof that the source was (in cross section) a rotor. Despite these caveats it is difficult to imagine any other ring-like source of excitation that could suddenly appear in the medium as a result of such stimulation and then repeat at the period typical of reentry (rotors), unless it were a ring of pivots: a filament of phase singularity.

The question is sure to be resolved soon, given the increasing refinement of epicardial, endocardial, and intramural mapping experiments, now almost to the ultimate unit of continuity, the 500-1000 micron space constant of healthy cardiac muscle (see Box 5.C). Moreover, by patterned illumination of tissue infused with photosensitive ionophores, it may become feasible to stimulate in a more controlled way than Medvinsky et al. managed with single electrodes. Rotors thus created on a surface must somehow connect three-dimensionally through the unilluminated depths. Though contemporary measurements do not yet have adequate resolution to discern their shapes, wavefronts in heart muscle are inevitably three-dimensional [Durrer et al. 1970; Brusca and Rosettani 1973; Spielman et al. 1978; de Bakker et al. 1979; Janse et al. 1980; Cohn et

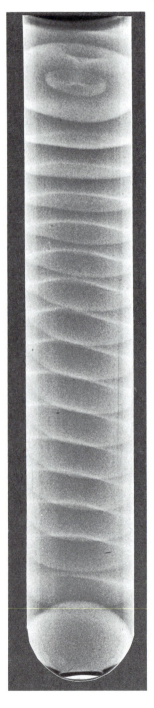

Figure 8.14: Sixteen cycles of activity in the scroll ring at the top of this 10-mm test tube have created as many concentric shells of excitation propagating away. Disk-like segments of these shells are still visible, propagating toward the bottom of the tube. This scroll ring is upside down relative to all figures except 8.9. From Welsh 1984, with permission.

al. 1982; El-Sherif et al. 1982; Medvinsky and Pertsov 1982; Wit et al. 1982; Spach and Kootsey 1983; Downar et al. 1984; Ideker et al. 1984; Medvinsky et al. 1984; Witkowski and Corr 1984; Kramer et al. 1985]. Three-dimensional reentry is sometimes observed (e.g. El-Sherif et al. 1985) around cores of functional conduction block as small as 3 mm in diameter; when their anatomy is fully resolved, will we see scroll rings?

Both of the foregoing demonstrations depended on reconstruction of evidence obtained piecemeal. Wouldn't it be more convincing simply to see a fully three-dimensional scroll ring rhythmically pulsing like a tiny jellyfish in a pond of nutrient water? By peering into the malonic acid reagent with a stereo microscope, a trained observer can recognize a scroll ring in projection as a curiously periodic pattern of red and blue. It repeats at the standard rotation period of planar spirals: no surprise, since it is made of planar spirals, laterally continued into the third dimension.

For a decade, nonetheless, only a handful of "trained observers" were convinced, because the perceptual advantages of color, of stereo vision, and of movement are all lost in the retelling. But by optimizing the recipe, the lighting, and the choice of a color video camera, B. J. Welsh in Glasgow persuasively documented scrolls and scroll rings as part of his Ph.D. research in mathematics [Welsh et al. 1983; Welsh 1984]. He also employed conventional photography to good effect, as shown in Figures 8.13 (cf. 8.7) and 8.14 (cf. 8.15).

The scroll ring is a little chemical clock, a periodic machine with no moving parts. Being chemical and liquid, it looks almost physiological. At regular intervals it creates within itself on a little "umbilical" cord a new and smaller replica (Figures 8.14, 8.16, cf. Figure 8.17). The cord breaks, freeing the little one to repeat the same cycle and releasing the parent as a closed shell that is being thrown off like the periodically molted skin of a snake. The skin radiates away into the sustaining pond of energy-rich medium that would otherwise be inertly quiescent. As long as the organizing center pulsates in this pond, every droplet of the medium—except along the singularity, where all remains arrhythmic—is compelled to breathe red/blue/red/blue at strictly regular intervals, quickly metabolizing malonic acid to bubbles of carbon dioxide. When either the air (bromate oxidizer) or the organic food (malonic acid) runs out, the waves fail to propagate and the organizing center dies: all reverts to quiescence. Actually, it may die even sooner if the energy supply lasts long enough: scroll rings tend to contract by one percent or so (depending on curvature) with each rotation. When the ring has contracted to about two core diameters it extinguishes spontaneously like the last spark of a shrinking fire [Welsh et al. 1983; Welsh 1984; Panfilov and Pertsov 1984;

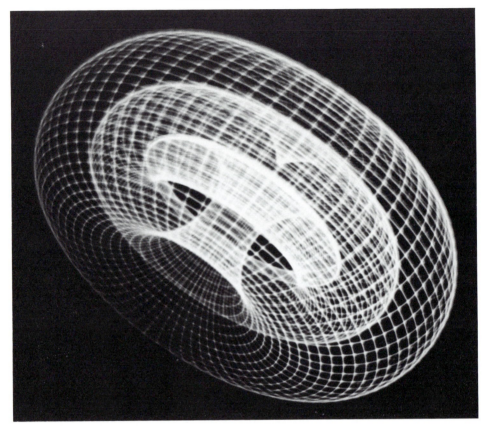

Figure 8.15: Figure 8.7 from a different viewpoint without cutaway: a complete symmetric scroll ring similar to Figure 8.14. From Winfree 1984b.

Panfilov and Winfree 1985]. This curious instability was not encountered in two dimensions. It depends upon the curvature of the scroll filament, a strictly three-dimensional concept [Yakushevich 1984].

But until shrinkage runs its course, the scroll ring can only be killed by violence, for example, by a shock that momentarily synchronizes the whole volume (the exact analogue of defibrillation by electroshock). Some additional modes of three-dimensional death can be understood by thinking for a moment of comparable processes in the plane. For example, suppose the two-dimensional medium were made inexcitable along a narrow corridor reaching from one rotor to another, counter-rotating. The winding number of phase along the border of that zone overlapping both singularities is zero. If inhibition is maintained long enough for all the paired wavefronts along the perimeter to destroy one another in

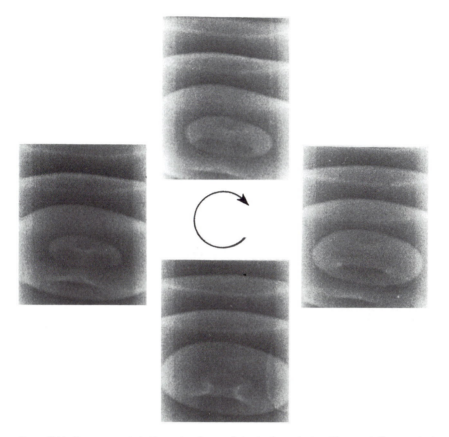

Figure 8.16: Four moments in the cycle of a scroll ring in the malonic acid reagent. Ten-mm test tube, as before, from Welsh 1984 with permission.

collisions, then the inhibition can be released without creating new pairs. In the simplest instance, the corridor is so short that it cuts across no wavefronts: two mirror-image rotors simply fuse and vanish. The presence of a nearby wall also catalyzes the extinction of a rotor, essentially by providing a virtual mirror image. Thus confinement in too small a box is eventually lethal.

These principles apply in three-dimensional context as well, except only that the "corridor reaching from one rotor to another" now spans the ring or connects it to another ring with appropriate orientation. Pairing of rotors in the plane becomes in three dimensions merely arcing across a single ring. A ring singularity can thus be destroyed by homogenizing only a pancake-shaped volume enclosing it, not necessarily the whole medium.

Figure 8.17: Computer-generated picture of the waves surrounding a scroll ring, just at the moment when the "umbilicus" is breaking to release another closed shell of wavefront. The inner unbroken continuous surface of wavefront, from break point to singular ring, is shaded. A wedge of about 90 degrees is made invisible in order to reveal wave interiors.

Though it lacks anything like a genetic system through which it could mutate and evolve, the organizing center shares many of the features that make living organisms interesting: chemical metabolism (oxidation of organic acids to carbon dioxide), self-organizing structure, rhythmic activity, dynamic stability within limits, irreversible dissolution beyond those limits, and a natural lifespan.

We have dealt with only Ideas 1 and 2 of the agenda above. That is where the story ends, as far as laboratory verification is concerned. But science is not just about redescribing more gracefully things already known. It is also about discovering new things. Many phenomena described in this book were discovered in experiments expressly designed to check a theoretical anticipation. In the next chapter we again anticipate, this time on the basis of Ideas 3 and 4.

CHAPTER 9

A Bestiary of Organizing Centers

Art is the lie that helps us to see the truth.
—Pablo Picasso

In Chapter 8 we found that the two-dimensional rotor of Chapters 5–7 is really a slice through a three-dimensional organizing center called a "scroll ring." This is a timely observation. With the advent of affordable microelectronics, cardiology has swiftly attained such electronic sophistication that even while they are deciding how to remove an arrhythmic focus in the operating room, surgeons can refer to a computer console for contour maps of electrical activity spreading across the surface of the exposed human heart [Downar et al. 1984; Figure 5.5 above]. But the recent literature of such advances repeatedly bemoans the fact that surface maps can be difficult to interpret, often in ways that suggest three-dimensional patterns of circulation below. The human left ventricle in particular (Figure 9.1) is thick enough to harbor organizing centers at least as intricate as any recognizable in the ideally uniform excitable media that lend themselves to three-dimensional observation. Do the observations of Chapter 8 constitute the last word on this subject? Are all forms of three-dimensional reentry equivalent to distorted versions or broken pieces of a scroll ring like Figure 9.2? Just what are the geometrical options available to waves circulating about pivots of conduction block? We invoke Polya again: "Good problems and mushrooms of certain kinds have something in common: they grow in clusters. Having found one, you should look around: there is a good chance that there are some more quite near." Let us consider some possibilities.

When looking for a new phenomenon it is often helpful to know in advance what to look for. We want only an artful outline, a caricature to tune our senses to the broad pattern of what we seek. In detail, the artistic simplification is of course a lie, but it can help us to see something

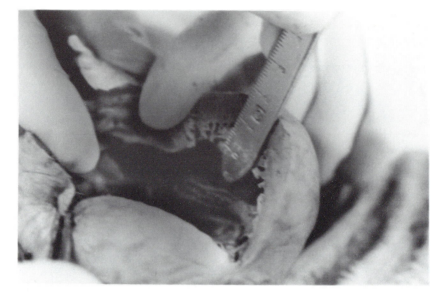

Figure 9.1: The ventricular muscle in an adult human is about 1 cm thick: space enough for reentrant circulation between epi- and endocardial surfaces. Electrophysiological recording in 3D lies even beyond present-day ambitious hopes. But it will eventually be necessary because 2D isochronal maps resist interpretation in terms of purely 2D propagation.

new for the first time. This chapter depends on an extreme simplification. We consider a three-dimensional volume in which something—it doesn't matter what—is happening periodically in time. Each point in the volume is like a tiny clock, whether because the full mechanism of rhythmicity resides everywhere (an oscillating reaction) or because each point is driven to rhythmicity from a remote periodic source (e.g. a wave-emitting distant rotor). We will not follow this cycle in time but will deal only with a "snapshot" of the volume, unconcerned whether it is passive, excitable, or spontaneously oscillatory. All we care about is the three-dimensional pattern of clock phases in that snapshot.

We begin to understand the variety of possible patterns by returning to the list of four basic principles, the first two of which were explored in the previous chapter.

Idea 3: Mutual Linking of Two Scroll Rings

Do scroll rings have social behavior? Are there other species of organizing centers? An affirmative answer comes from an unexpected direction: the discovery that more complex organizing centers can exist, composed of

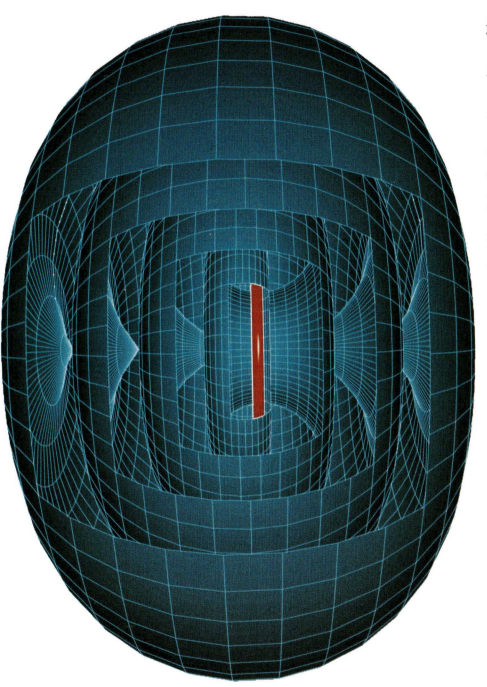

Figure 9.2: A scroll ring starting its fourth cycle would appear from outside as an egg-shaped wavefront. Here windows have been cut to reveal an arc of the singular ring (horizontal red) and the spiral cross section of waves emitted from it. Waves collide internally along the vertical axis of rotational symmetry.

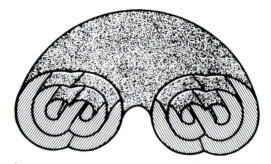

Figure 9.3: A pair of adjacent mirror-image rotors surround themselves in closed rings of wave-front. By duplicating this structure in all vertical planes through the vertical axle a pair of concentric scroll rings is generated, surrounded by toroidal surfaces of wavefront that collide along the axle.

suitably interlinked aggregates of individual scroll rings—but scroll rings qualitatively altered by their participation in the greater unit.

The simplest example requires no mathematical theory: just look at Figure 7.6 or 7.8. Here we see a mirror pair of rotors, counter-rotating about their respective singularities. Each emits a spiral wave. The waves collide, enveloping the pair in concentric shells of "skin." Figure 9.2 has this structure in every plane through the vertical axis. Figure 9.3 does also, but in a different way. Here we see a pair of concentric scroll rings. The waves emitted by the pair collide in an orderly way along the vertical axle of rotation of the whole structure. This innocuous elaboration on the foregoing theme now lends itself to a peculiar mutation: suppose we let the pair of rotors revolve once around one another as our viewpoint moves around the axle (Figure 9.4)? Nothing is qualitatively altered outside a doughnut-shaped region delimited by the first concentric shell of "skin." Inside that shell something is altered, but subtly. The alteration is scarcely detectable locally. In any little wedge of doughnut, we still have a pair of stubby scrolls counter-rotating next to one another. The fact that the line joining them turns imperceptibly from one wedge to the next only mildly distorts the previous pattern of chemical concentrations.

But regarded globally, not just locally, we recognize this distortion as a qualitative change. The two singular rings are now linked! Not only are they linked, but each is also twisted topologically in a way that necessitates linkage to its twin.

Idea 4: Scroll Rings Differ in the Number of Twists Locked into the Ring

The simplest way to perceive this twist is to concentrate attention on one of the twinned pair of linked rings. Imagine a smaller doughnut enclosing

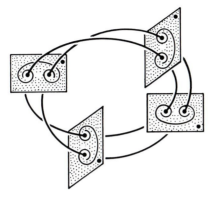

Figure 9.4: As in Figure 9.3, except that the pair of singularities rotate around one another as they orbit around the axle. The two scroll rings are now linked. Moreover, each is a once right-twisted scroll. Compare with computation in Figure 9.14. From Panfilov and Winfree 1985.

only that one. The scroll inside that tube rotates once as we follow it fully around the ring in a snapshot. It is twisted just as though we had made it by taking a plain scroll ring, opening it straight, imposing a full turn on one end of the cylinder, then bending it again into a ring and sealing it shut again (Figure 9.5, bottom). This twisted scroll is different. For example, as Figure 9.5 (top) shows, the untwisted scroll emerges from its enclosing doughnut along a horizontal ring (painted darker for phase zero); but the twisted scroll emerges along a ring that links through the hole in the doughnut and links the singular ring. But the most telling aspect of its distinctness is revealed along the inside equator of the doughnut.

In Figure 9.6 we color-code the local phase on snapshots of the doughnut surface containing, above, a plain, rotationally symmetric scroll wave and, below, a twisted scroll wave. Notice that unlike the rotationally symmetric case, the bottom structure has a complete sequence of all colors around every equator.[1] Do you have a feeling of déjà vu? We have been here before, and we must recognize the same peculiarity about this situation: somewhere in any disk bounded only by that ring there must be a colorless point. The volume outside the doughnut thus cannot be filled with a continuous field of color (a continuous arrangement of periodic

[1] Because there is no unique sense of "clockwise" or "counterclockwise" in three dimensions, this could be called "winding number = 1" or "−1," arbitrarily. However, the "twist" is distinguishably either +1 ("left") or −1 ("right"). The necessary conventions are sorted out in the appendices to Winfree and Strogatz 1983bc. From 1980 to the end of 1985, my papers illustrated left twist in the linked pair of scroll rings except for an accident in Figure 12 of Winfree 1984b. Right twists are shown throughout this chapter and, except for accidents in Figures 1 and 5, in Panfilov and Winfree 1985. Knots have been portrayed as left- or right-twisted at random.

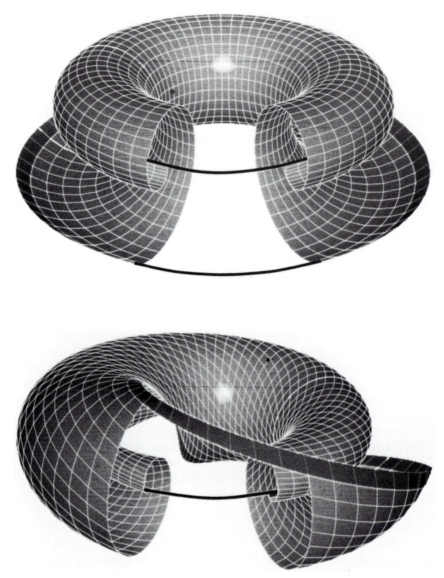

Figure 9.5: Top: An untwisted scroll is confined within an invisible torus. Where it protrudes (only one cell-width), it is painted darker grey. The foreground segment is sectored open for visibility, revealing an arc of the horizontal singular ring. Bottom: The same picture is redrawn but with the spiral wave gradually rotated as it is carried around the vertical axle. This results in a twisted scroll wave which exits the torus along a linking ring (darker grey). Adapted from Winfree 1984b with permission.

timing, a continuous wavefront). In fact the doughnut must be threaded by another, unforeseen singular filament!

You can see the problem physically. Suppose the twisted scroll ring of Figure 9.5 (bottom) were continued beyond its confinement to the inside of a doughnut. The spirals drawn in every radial plane, centered on every point along the source ring, collide at the vertical axle of the doughnut. But they do not collide synchronously and symmetrically as in Figures 8.7, 8.15, and 8.17. Figure 9.7 shows how they do collide: the merits of keeping the solitary twisted scroll quarantined within a doughnut can thus be appreciated! Waves outside the doughnut arrive at the axle in succession from sources arrayed around the doughnut equator: they arrive as an inward-propagating scroll, and their collision, as we first saw in color codes for timing, is its axial phase singularity.

This simple extrapolation of the scroll ring thus seems at first to require an additional phase singularity of the wrong kind: it has the right winding number, but it is a sink rather than a source for waves. To be "realistic," perhaps it should be replaced by the only known singular filament, the scroll source. That outward scroll wave would collide (along an interface between the two sources) with the wave incoming from the twisted horizontal scroll. Figure 9.8 re-presents the horizontal right-twisted scroll wave of Figure 9.7 but shows it only as far as a collision interface which, in this picture, is close to the vertical axial singularity (and hides it).

Of course, we already knew that a twisted scroll ring may (we just didn't know that it must) have a partner: the other is the twin, the linking scroll ring of Figures 9.3 and 9.4. The lesson to take home from the topological argument is that if some non-zero integer twist is locked into scroll ring, then it must be linked by a twin; and conversely, if a solitary pair of rings are linked, they must each be appropriately twisted. The number of full-cycle twists (suitably defined) turns out to just equal the algebraic sum of linkings by clockwise and counterclockwise scrolls. This "suitable definition" was discovered by a curiously circuitous route.

Quantum Numbers for Organizing Centers

The roads by which men arrive at their insights into (scientific) matters seem to me almost as worthy of wonder as those matters themselves.

—Johannes Kepler,
Astronomia Nova (1609)

One of the prime mysteries about the genetic material, DNA, is its physical arrangement in the nucleus of every cell. DNA is a chemically oriented filament—actually a ribbon whose edges are the two oppositely oriented complementary filaments. It has a mechanical bending moment, torsion,

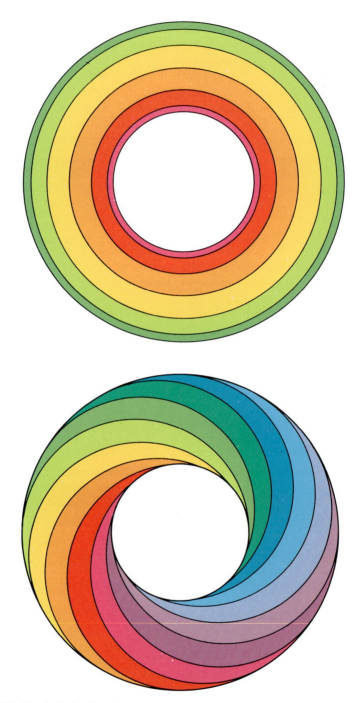

Figure 9.6: Top: Loci of uniform phase are painted on the surface of a torus confining an untwisted scroll ring. Bottom: As above, but the scroll ring is twisted: around the inner equator of the torus, all colors appear in sequence.

224

Figure 9.7: Figure 9.5 (bottom) is continued beyond the confining torus. Wavefronts converge to a phase singularity along the vertical axle. From Panifilov and Winfree 1985 with permission.

and so on, like any other physical ribbon. Many meters of it must be somehow folded, wound, pleated, knit, woven or otherwise packed into the microscopic confines of every living cell: a volume so small that at least a million 180° turns are required. Moreover, during every cycle of cell division, this compact bundle must unravel to expose every bit of itself to reading enzymes. Then the two strands must separate to copy themselves. The whole incredible length must completely separate into two distinct bundles without becoming hopelessly entangled and knotted.

How is this managed? In short, the strands do break and rejoin, but the details are only now being found out experimentally. Persistent theoretical inquiry has led to the elaboration of a mathematics about twisted, writhing ribbons, often closed in mutually linked rings. Theorems about DNA coiling and supercoiling emerged during the 1970s and have since matured to a concern with knotting as well [a short bibliography: Delbruck 1962; Fuller 1971; Crick 1976; Fuller 1978; Bauer et al. 1979; Wang 1982; Wasserman et al. 1985].

As an undergraduate mathematics major, Steven Strogatz gave thought to this knotty chapter of cell biology [Strogatz 1980; Strogatz 1983].

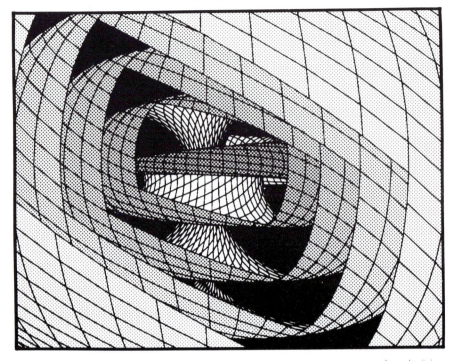

Figure 9.8: As in Figure 9.7, but instead of sectoring the organizing center open to the axle, it is windowed with parallelograms. The axial singularity is not visible.

Later encountering the subject of linked and knotted singular filaments, he developed the analogy far enough to formalize the twisting and writhing of singular filaments in organizing centers [Winfree and Strogatz 1983bc]. That development can be told briefly as follows. Prominent among the pertinent equations of differential geometry are two complicated integrals based on the shape of a ribbon of wave-edge. One, called Tw, the twisting number, integrates the twisting of that ribbon. The other, called Wr, the writhing number, depends only on the way the singular filament departs from a plane. Neither one is usually an integer, but their sum L, the twist integer, is, if the filament closes in a ring. That closure naturally quantizes the description of organizing centers in terms of integers (of course, not involving Planck's constant). As it turns out [Winfree and Strogatz 1984b; Winfree et al. 1985], this so simplifies the whole subject that the differential geometry quietly neutralizes itself and disappears. Having thus systematized the idea of integer quantum numbers, the initial mathematical structure proves to have been a catalytic scaffolding that is completely dispensable.

In the end, the central construct is a device invented by topologist Herbert Seifert for fitting a smooth surface—a "Seifert surface"—to any arrangement of rings [Rolfsen 1976]. That surface provides a way to visualize the surfaces of uniform chemical concentration that collectively constitute the wave field. The anatomy of organizing centers then reduces to just four principles:

(1) Any arrangement of mutual linkage among any number of scroll rings is permissible. There is always a way (Seifert's) to construct a wavefront that fits all the ring singularities as edges;

(2) Looking at this wavefront close to any one ring, we find its edge to link some integer number of times around the ring: that integer, L, can be shown to be the sum of Tw and Wr integrals or, more simply, the total winding number of phase associated with any linking filaments. It can be evaluated by counting ($+$ or $-$, according to direction) punctures of the given ring's Seifert surface by other rings. This is the appropriate definition of "twist" referred to above, rather than the Tw (twisting number) of differential geometry;

(3) Knotting, though it ostensibly complicates the picture and provides additional variety of topological isomers, does not affect an organizing center's linkage relationships; and any arrangement of tame knots is admissible. Thus, having contributed the Seifert method, the theory of knots proves irrelevant after all;

(4) Organizing centers big enough so that their singular filaments are locally almost straight and untwisted emit waves at the common period of the two-dimensional rotor. The waves collide to form concentric sphere-like bags surrounding the whole organizing center.

What about the opposite limit, in which scroll ring diameter is comparable to the diameter of a rotor? This happens when proportions are small or diffusion is rapid or kinetics are slow: three ways of verbalizing the same limit. In such organizing centers, diffusion, twist, and curvature may become the dominant considerations. We currently know nothing of their rotation speeds, shrinkage rates, and stability except from a few very large and expensive computer simulations. It appears that the local behavior of a rotor (a segment of singular filament) in three dimensions differs from its bland two-dimensional behavior in two respects:

(1) The filament may be curved, in which case it moves very slowly toward the center of curvature [Yakushevich 1984]. This movement tends to straighten an open filament and to make a closed filament exactly circular. The circle continues to shrink and, as it shrinks, to drift slowly (a few percent of propagation speed) in the direction of wave propagation through the interior [Panfilov and Pertsov 1984; Reiner and Winfree 1985].

(2) The filament may be twisted, in which case the more twisted part rotates somewhat faster than the rest. On an open filament, this alters the distribution of twist. In a closed filament the total twist is conserved, so the period is stably shortened [Panfilov and Winfree 1985]. These reaction-diffusion computations show that Figures 9.7 and 9.8 are more realistic than we originally realized. If the unclosed vertical axle of singular filament has no twist, then it emits spiral waves at the standard frequency of a rotor; but the twisted horizontal ring emits its waves at higher frequency, so they collide nearer and nearer to the axial rotor, eventually squeezing it out of existence as in Figure 9.7.

Thus rotors are capable of more subtle quantitative variation in three dimensions than in two. Nonetheless, from merely topological principles, derived circuitously by way of the theory of knots and the differential geometry of DNA, we can determine which organizing centers are mere figments of a diseased imagination (see Box 9.A) and which others are chemically realizable. It still seems too difficult to derive their long-term behavior, due to the subtle movements and changes of period induced by curvature and twist of the singular filament. But given the quantum numbers of a viable organizing center, can we at least specify what initial arrangement of chemicals in space will evolve into a representative of that topological type?

In short, yes. The initial conditions are just those needed to initiate two-dimensional rotors, projected along the singular filaments in three dimensions. They can be fashioned from chemical gradients [Winfree 1985; Poston and Winfree 1987] or from fragments of preexisting waves [Panfilov et al. 1985; Panfilov and Winfree 1985; Winfree 1984b; Winfree et al. 1985]. Their topology can be made particularly transparent: the needed fragment is a Seifert surface for the intended configuration of rings. Thus a rotationally symmetric scroll ring evolves from a disk of wavefront: its free edge becomes the singular axis. A linked pair of once-twisted scroll rings evolves from a twice-twisted band of wavefront (see Box 9.B). A solitary knotted scroll ring evolves from a pair of disks joined by three half-twisted bands (Figure 9.13). Such initial conditions are conceptually transparent, they are perfectly suited for computation in discrete-state discrete media, and they will serve "in a pinch" for computation in continuous media. But they constitute only a poor caricature of the smooth concentration gradients actually wanted when thinking realistically of chemical organizing centers (see Box 9.C). Such smooth patterns can be described very simply by four-dimensional algebraic polynomials, mutating integer coefficients and exponents to initiate diverse arrangements of linked, twisted, and knotted rings (see Appendix).

Returning to the solitary pair, we might think of Figures 9.7 and 9.8 as drawings of an extreme case in which one of the pair is the horizontal ring contained within the field of view, but the other ring source is very large and passes through the field of view only as a vertical segment. The incoming wave from the horizontal source suffers its inevitable collision with the scroll radiating from the vertical central segment of the big ring, quite near to the latter. Seen thus as a limiting case, Figure 9.8 has the merit of displaying the entire anatomy of one right-twisted scroll ring unobscured by the other. Jagannathan Gomatam of Glasgow College of Technology was able to show analytically that something similar is indeed a solution to the linearized equations of reaction and molecular diffusion [Gomatam 1982]. It has the supplementary merit of lending itself quite conveniently to animation.

Box 9.A: What Organizing Centers Can (Cannot) Exist?

How often have I said to you that when you have eliminated the impossible, whatever remains, however improbable, must be the truth?
—Sherlock Holmes, in "The Adventure of the Sign of Four"

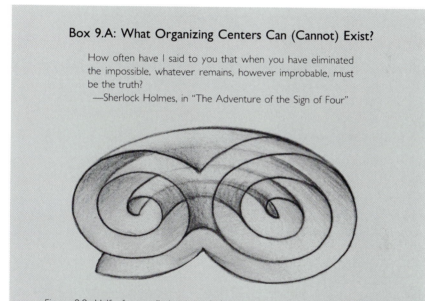

Figure 9.9: Half of a scroll ring intended to exhibit "exotic topology," drawn by hand. As it turned out, this is no surface at all, but only an optical illusion. From Winfree and Strogatz 1983b with permission.

Trying to draw nontrivial organizing centers is hopeless. Figure 9.9, for example, is one of my better attempts, vintage 1980: if you study it carefully you will be reminded of Escher's impossible objects. Even computer graphics can go astray: Figure 9.10 was supposed to fit together two twisted scroll rings in a physically realistic way; but there was an error in the code, overlooked only because the impossible mismatches are hidden. One needs a theorem. An educated guess based on winding numbers, differential geometry, and physical chemistry first produced it [Winfree and Strogatz 1984b]. The proof [Winfree et al. 1985] turns out to require nothing more than pure topology. The index theorem about linked singularities:

> Each ring's twist integer L must be equal and opposite to the sum of the positive and negative integers that describe its linkage with all the other rings. The "twist" here is understood as the mutual linkage L of the two edges of a thin band of wavefront cut adjacent to the singular ring—*not* as the Tw of differential geometry referred to in the text.

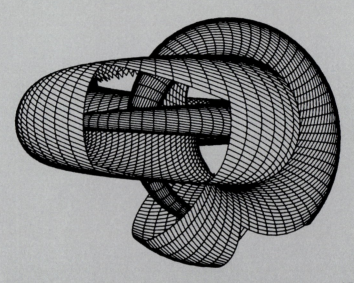

Figure 9.10: A pair of identical right-twisted scroll rings (Figures 9.5(bottom), 9.7, 9.8) with windowing for visibility are fitted together to link perpendicularly. This computer graphics attempt (from Winfree 1984b, Figure 18, with permission) is another mistake: one ring is upside down and the two wave surfaces cannot join smoothly.

Proof:

Phase varies continuously with position everywhere except at rotor pivots. Pivot points are arranged along singular filaments, which generically form closed rings. A single (scroll-shaped) wavefront extends from each ring, joining with the wavefront of other rings where waves collide, to form a single surface bounded only by the rings.

To convince yourself that there can be a single non-self-intersecting surface whose boundary is an arbitrary set of linked and knotted rings, construct it thus:

(1) give each ring a direction by putting an arrowhead on it;
(2) project the collection of rings onto a wall and note where their shadows cross;
(3) erase each cross and reconnect the dangling ends in the "short-cut" way that respects direction without crossing;
(4) you now have a lot of disjoint flat rings: fill each with a disk;
(5) where rings are nested, these disks are stacked: separate them by pulling the innermost forward off the wall, and so on;
(6) go back to each erased cross-over and join those two disks by a half-twist band so that its two edges cross in projection on the wall as did the erased segments.

Back to the proof: Paint the two sides of this wavefront surface red and blue, then close it by capping each ring in its boundary, and extend the paint job to these caps. These caps are disks; they generally intersect the red-and-blue wavefront surface and each other; some of them (e.g. on any knotted ring) may be self-intersecting. That is o.k. Call the capped "wavefront surface" the "closed surface."

Now consider any closed curve lying on the wavefront surface adjacent to one of the rings, oriented in the same direction. (This is the other edge of the thin band of wavefront alluded to above.) We will evaluate its "intersection number" with the closed surface. The intersection number of an oriented closed curve with an oriented surface is the number of times the curve penetrates from red to blue, minus the number of penetrations from blue to red. Thus it is zero for any closed surface. The closed surface is the wavefront plus the caps. The intersection number with the wavefront is zero, since the curve lies on the wavefront. So the intersection number with the collection of caps must be zero. The intersection number with each cap is the linking number of that cap's boundary ring with this curve: in the case of its own cap, it is the negative of the twist number, L; in the case of each other cap, it is the same as the linking number of this curve's adjacent boundary ring with that cap's. Thus L must equal the sum of linking numbers, and this must be true for each ring.

This resulting index theorem seems so simple. Does it exclude anything other than the obvious monstrosities depicted above? Yes, indeed it does. It is not otherwise perfectly obvious, for example, that a solitary pair of plain scroll rings without twist can mutually entangle as in Figure 9.11a, but, because each must have net linkage = 0, cannot link as in Figure 9.11b.

This index theorem and exclusion principle also says something still less obvious. Suppose a singular ring could split in two, or two could fuse into

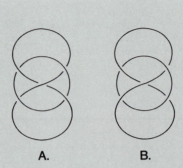

A.　　　　**B.**

Figure 9.11: (a) The Whitehead link: two untwisted singular rings are mutually entangled: they cannot be separated without cutting one. But they are not linked in the mathematical sense: draw arrows in each ring and observe that each oriented filament penetrates any oriented disk spanned by the other an equal number of times in opposite directions. (b) Two singular rings can link one another twice; each scroll then must be twice-twisted.

one. (A mechanism for such transmutation is imaginable; see Winfree and Strogatz 1984a.) If the exclusion principle must be satisfied before and after, the possible transmutations are constrained. For example, if the exclusion principle had turned out to allow only organizing centers with an odd number of rings, it would have forbidden all such transmutations. As it is, it only requires that the fused ring link each other ring as many times as the separate two did together, and that its own twist L is the sum of the two rings' plus twice their mutual linkage. (This is obviously satisfied in the two trivial transmutations—a single scroll ring splitting into two separate ones, and one vanishing—that have since been observed [Welsh 1984].)

The index theorem provides a taxonomy of the possible organizing centers in three-dimensional periodically active media and excludes the much greater diversity of imaginary others. It does *not*, however, guarantee stability of the topologically admissible organizing centers: there may be yet other constraints that further winnow the possibilities for persistent organizing centers. At this writing no other constraints are known, so stability can only be checked numerically. Such computations have been carried out in approximation for the solitary scroll ring, for the linked pair, and for the solitary trefoil knot.

Box 9.B: Computer Simulation of Excitable Media in Three Dimensions

There has been an alarming increase in the number of things I
know nothing about.

— Ashley Brilliant
Potshot #729

In Figure 5.2 we took a quick look at a game played with cellular automata
to simulate an excitable medium. Each cell could be excited, refractory,
or susceptible. Starting with arbitrary initial conditions, waves propagate,
usually in the form of some combination of clockwise and anti-clockwise

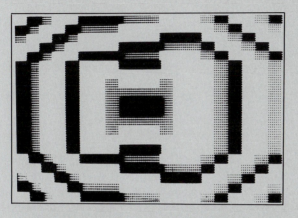

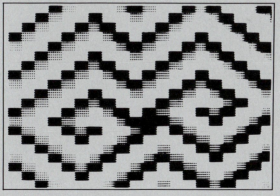

Figure 9.12: Perpendicular slices through a single untwisted scroll ring propagating
waves at period 4 in a three-dimensional array of 8000 excitable cells. Black cells
are newly excited; grey are refractory. Compare continuous version in Figures 8.9
and 8.10.

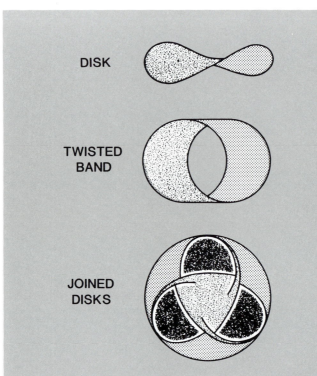

DISK

**TWISTED
BAND**

**JOINED
DISKS**

Figure 9.13: Sketched initial conditions for (top) an untwisted scroll ring, (middle) a pair of linked right-twisted scroll rings, and (bottom) a solitary trefoil-knotted scroll ring. The singular filament forms along the edge of a patch of wavefront. Diagonal shading represents the active front, stipple indicates the refractory wake.

spirals. Played on quadrille paper as here, instead of on a hexagonal tiling as there, and allowing only one time-step in the excited or refractory states, rotors turn with period 4 (just above the lower limit of 3 set by the refractory period). This game lends itself to numerical simulation. Iteration of these rules on a 20 by 20 grid of squares, for example, goes so fast, even on an old Apple II computer, that extension from two dimensions to three immediately suggests itself. My son Erik implemented this suggestion one morning during high-school summer vacation on his home-built computer. A scroll ring is easily initiated by placing a square of excited cells, perhaps 10 by 10, inside a 20 by 20 by 20 cube, with a square of refractory cells adjacent to give the wavefront a direction. Within a few hours its edge curls up to become a scroll ring, as shown in these serial sections (Figure 9.12). Fancier two-layer initial conditions initiate topologically distinctive organizing centers (Figure 9.13). Using twisted-band initial conditions as in Figure 9.13 (middle), the first linked pair of

Figure 9.14: Perpendicular slices through a linked pair of right-twisted scroll rings in a three-dimensional array of 64,000 excitable cells. Waves are propagating at period 6 (because the model used here has two more refractory states than in the 3-state model used above). In each plane you can see four phase singularities where the two rings are each encountered twice. To assemble three-dimensionally, rotate the top picture out of the page plane by pushing its left edge 90° behind the page while lifting its right edge equally above the page plane; then slide it parallel to the margin of the page down through the bottom picture, so that each cuts through column 20 of the other. They agree along this intersection. The top right and bottom right pairs of mirror-image spirals correspond to those in the top right and bottom right of the anticipatory schematic in Figure 9.4; top left here is rear left in Figure 9.4; and bottom left is leftmost in Figure 9.4. From Panfilov and Winfree 1985 with permission.

twisted scroll rings (Figure 9.14) was computed (after moving the PASCAL program to a COMPAQ DESKPRO with capacity to store 51^3 integers). The initial conditions of Figure 9.13 (bottom) gave rise to a knotted scroll ring. All proved stable in this kind of simulation.

How can we implement the same geometries in a chemically more realistic way? A step in that direction is achieved by finely dividing space into an arbitrary number of tiny volume elements ("cells") and assigning to each not one of a small set of discrete states, but two (or more) chemical concentrations. The concentrations are then allowed to change according to differential equations of synthesis and degradation supplemented by molecular diffusion from adjacent volume elements. The first such computations were produced in Puschino, USSR, in a grid of 27,000 volume elements [Panfilov and Pertsov 1984; Panfilov and Winfree 1985; Medvinsky et al. 1984; Panfilov et al. 1984; Panfilov et al. 1985]. Given efficient programming of simple kinetics and a $50 \times 50 \times 50$ reticulation of space, a scroll ring can be followed through several rotations overnight in a COMPAQ DESKPRO, for example [Reiner and Winfree 1985]. Using the L_{ij} quantum integers in appropriate complex-valued polynomials for initial

235

conditions (see Appendix), the period and stability of various distinct organizing centers can be evaluated numerically. This exercise will have particular interest when simplified kinetic schemes for generic excitable media are supplanted by: (a) the Oregonator kinetics for the Belousov-Zhabotinsky reaction; and (b) the Beeler and Reuter [1977] kinetics for ventricular muscle, possibly corrected by the Ebihara and Johnson [1980] wavefront kinetics (or possibly not if the simulation should mimic "slow response" reentry in depressed tissue).

Results might be useful in guiding attempts to initiate diverse organizing centers in chemical media and in heart muscle, and for recognizing them if they should appear spontaneously or in response to such deliberate interventions.

Box 9.C: Why Should We Use Continuous Models?

How do discrete-time, discrete-space excitable media differ in their behavior from continuous excitable media? One way to put it is that the behavior of waves in such media constitute by definition the "trivial" aspects of wave behavior in excitable media. The behavior of waves in continuous excitable media may be thought of as consisting of the "trivial" behavior exhibited in these simple models, modified in ways that depend on the exact nature of the particular continuum.

One of the simplifications that characterizes simple discrete-time discrete-space excitable media is that adjacent cells are coupled by a rule that bears little resemblance to molecular diffusion or the spread of electric currents (by the same mathematical rule in both cases: the Laplacian operator). Even though it improves the representation of space and time from discrete to continuous, the Wiener and Rosenblueth [1946] model of cardiac muscle, for example, still fails badly in some respects because it uses the coupling rule used in such discrete-time, discrete-space models, rather than using the Laplacian.

Of course *all* models implemented in digital computers are necessarily discrete in time and space, yet appropriate digital models do successfully simulate continua. If there is thus no distinction in principle, where does the difference lie? It lies in the elaborateness of the rules. The important point is not the discreteness of the model's states, but in *how* discrete they are and what kind of rules govern the state transitions. Simulations intended to mimic continua assign to each discrete point in space anywhere from one to several state variables, each represented by a real number and thus capable of adopting a great many discrete states—at least several bits of storage are needed for each variable (perhaps a hundred states) for a representation that sensibly resembles a continuum. Regarded as a cellular automaton, the set of rules implemented to mimic a continuum is much more elaborate than the rules I have in mind when referring here to "discrete-state discrete-time" models of excitable media such as were deployed by Moe et al. [1964], Reshodko [1973], Reshodko and Bures [1975], Greenberg et al. [1978], Hastings [1981], Madore and Freedman [1983]. I will use the term "cellular automaton" henceforth to distinguish such minimal discrete-state discrete-time models.

In what way does their behavior oversimplify the real behavior of continuous excitable media? The answer depends on personal evaluation: is a given behavior "important" or "a detail"? Here are six behaviors in which differences are noticeable.

(1) Cellular automata propagate the impulse at unit speed regardless of wave number in periodic wave trains. There is no dispersion curve (or you might say there is, but it is identically flat and, while still at unit speed, ceases to exist at wave spacing equal to the number of discrete states defined). Continuous media exhibit idiosyncratic behavior depending on their dispersion curves.

(2) Two- or three-dimensional cellular automata also propagate the impulse at unit speed regardless of wavefront curvature. Thus they have no critical nucleation size for excitation. Or you might say that the smallest allowed nucleus of excitation—the unit cell—is already coarser than the critical nucleus. On account of this peculiarity, cellular automata also exhibit none of the other critical phenomena that depend on having an upper limit to allowable wavefront curvature (see Zykov and Morozova 1979)—or you might say the graininess of the medium is such that the greatest possible curvature is less than this limit anyway.

(3) Such media have no "core" or "rotor" at the source of spiral waves. The wavefront propagates with standard profile at unit speed everywhere and simply ends discontinuously along a "crack" (a plane, line, or point), the other side of which consists of refractory medium behind another normally propagating front. In contrast, the rotor in a continuum has a finite volume containing a chemical structure consisting of transverse concentration gradients, in which the pulse profile is different and "propagation speed" falls in proportion to radius as the whole structure rigidly rotates.

(4) Stability is an interesting issue in two-dimensional reaction/diffusion media, but it cannot be investigated in the cellular automaton analogue: its spiral waves are all stable. In diverse reaction/diffusion continua, in contrast, those with different numbers of arms may be stable or unstable and may spin at different rates depending on the number of arms [Krinsky and Malomed 1983; Malomed and Rudenko 1986]. In three dimensions, or-

ganizing centers slowly shrink in continuous media, but slow deformations cannot occur in discrete media.

(5) Rotors in continua do not always remain rigidly fixed where created as in cellular automata, but can meander along peculiar paths that remain to be investigated. Whether they meander or not apparently depends on a dimensionless ratio of certain parameters. The dependence of rotor diameter (and therefore stability in a confined medium) on those parameters is opposite on opposite sides of the critical ratio. The analogous dependence in cellular automata, with nonmeandering spirals, corresponds to the meandering spirals of continuous media [Zykov 1984].

(6) In parameter gradients or electric fields, rotors in continuous excitable media migrate at characteristic angles. There is no way to implement continuous gradients, e.g. of refractory time or of excitation threshold, in simple cellular automata.

Cellular automata contrived to resemble excitable media demonstrate in the simplest possible way the overall geometry of wave fields propagating away from their sources. That simplicity is their great beauty. But if we care to study the inner structure of those sources, their stability, their spontaneous meander, or their drift in gradients, or if we need to reckon with the consequences of high frequency, of curvature, or of exogenous stimuli, we unfortunately must come to grips with the continuum. We can do this using the rather elaborate cellular automata called a digital computer and its program to solve differential equations or, someday, we may do it analytically.

A Four-Dimensional Symmetry

Animation of the twisted scroll ring is accomplished trivially because this whole structure advances in time by rotating about the vertical axle. How can that be? By turning Figure 9.7 about its vertical singular axis, we move each radial wedge into the position formerly adopted by the slice just behind it—but rotated as that slice would have been had time advanced instead. This symmetry connects the three dimensions of space with the fourth dimension of time in this wave. I can think of no more fitting way to stitch together the pages of this book's exploration of cooperation, timelessness, and arrhythmia in biochemical clocks than by celebrating this microscopic machine made of phase singularities, whose moving parts are only bands of rhythmic color. Figure 9.15 shows a foreground cutaway presentation in black and white with background illumination dimmed.[2]

[2] Stereoscopic views of this "helium" organizing center in Winfree and Strogatz 1983b, Figure 40, and in Winfree 1984b, Figures 13 and 15, can only be seen by holding the page

Thus the anatomy of the twisted scroll ring lends itself to description as:

(1) an arrangement of waves in which every piece of wave surface (except so near the singular filament as to be within the rotor itself) propagates in the usual rectilinear way at a standard velocity from a phase singularity; and

(2) a necklace of two-dimensional rotating spirals, each turning on a circular axis; and

(3) a three-dimensional object of curious asymmetry rigidly rotating about its vertical axle.

However it may be regarded, this object, like the solitary untwisted scroll ring, is an organizing center for waves throughout the whole volume of excitable material. Those waves sequence the chemical reactions in time and lay out the timing in space as moving concentration gradients. Down these moving gradients materials flow from regions of momentarily excessive synthesis into regions of momentarily excessive degradation, integrating parochial reactions into a coherent organism.

Transmutation

We have seen two topologically distinct kinds of organizing center, a solitary scroll ring and a linked pair of twisted scroll rings. Topological differences usually cannot be mediated by any continuous process. It comes as a surprise, therefore, to find that organizing centers, nonetheless, are mutable. For example, scroll rings tend to contract slowly, and ultimately breathe forth one last nearly perfect spherical wave, then vanish [Winfree 1973bc; Winfree 1974d; Welsh et al. 1983]. See Figure 9.16. The same is observed in three-dimensional numerical solutions of reaction and diffusion [Panfilov and Pertsov 1984; Panfilov and Winfree 1985; Reiner and Winfree 1985].

Brian Welsh has also observed fission of a scroll ring into two (following which, his smaller one contracted and vanished) [Welsh 1984]. See Figure 9.17. That transformation in reverse allows for the creation of scroll rings *ex nihilo* and for the smooth fusion of tangent rings into one. The equivalent of fusion is commonly observed [Winfree 1973bc; Winfree 1974d] near an impermeable boundary when a scroll ring (and its twin in the mirror-image virtual concentration field behind the boundary) encounter

---

upside down: left and right images were accidentally transposed. You can see the left-twisted version in motion as a black-and-white videotape that is available for loan in VHS, Beta, or 3/4″ format (and by watching in a mirror you can see the isomer in which both scrolls are right-twisted).

Figure 9.15: In the same format as Figure 8.7, a segment of the wavefront of a twisted scroll ring encircles a near-vertical segment of a much larger twisted scroll ring. Waves from the smaller, nearly horizontal ring occupy the whole space, colliding at the vertical segment. The foreground has been sliced away, opening two elliptical holes in the wavefronts to enhance the view of the interior. What remains is drawn brighter in the foreground, dimmer to the rear. From Winfree and Strogatz 1983b with permission.

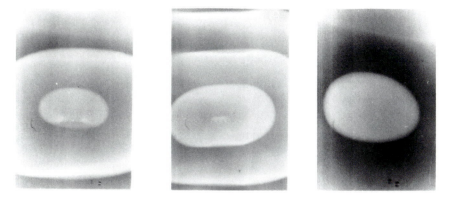

Figure 9.16: In the left panel a scroll ring periodically emits shells of oxidation. In the middle the ring has dwindled to precarious diameter. On the right, one last pulse radiates away from the site of the vanished singularity. From Welsh 1984, Figure 3.11, with permission.

one another: this is fusion "catalyzed" by the interface, which plays no dynamical role except for ensuring perfect alignment at tangency. Otherwise these may be rare events unless the medium is quite congested with a tangle of singular filaments. The analogous tangles of quantized vortex lines in superfluid helium behave in many ways like the quantized singular filaments in excitable media; and they are thought to reconnect upon arbitrary transversal encounter in three dimensions [Schwarz 1978; Schwarz 1982; Schwarz 1983; Schwarz 1985]. Could it be equally easy in excitable media? Whether commonly or rarely, such topo-chemical events can occur and the rules of permissible transmutation can be written as equations [Winfree and Strogatz 1984a; see Box 9.A]. Transmutations potentially allow any organizing center to change through a series of such events into any other. From another viewpoint this should be no surprise, because organizing centers inhabit chemical or biochemical media that will eventually lose their excitability and "die"; unless they just fade away like animals seen by twilight, every organizing center must have a pathway to extinction, presumably as a single shrinking ring in the end.

Vortex Atoms and Cosmic Strings

Though modern computer graphics lend a shiny new look to any discussion of organizing centers, the modern era is not the first to witness a speculative application of these ideas. Near the end of the American Civil War, half a century before the advent of quantum mechanics in Germany and Denmark, there emerged from the ingenious imaginations of Hermann von Helmholtz in Germany and Sir William Thomson (later

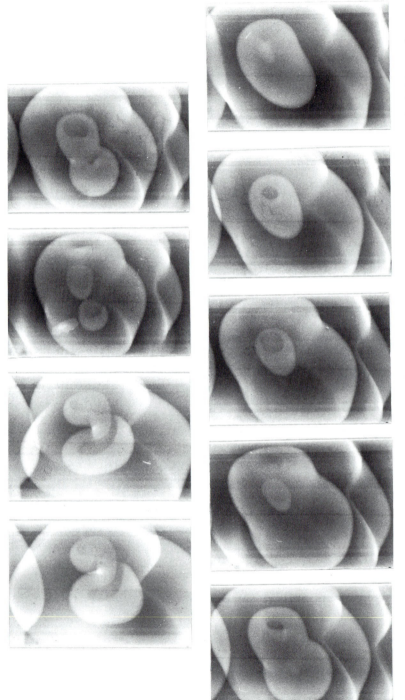

Figure 9.17: In nine successive photographs a single scroll ring in Belousov-Zhabotinsky reagent (in projection, twisted like a figure 8) divides into two smaller disjoint rings, one of which then dwindles and vanishes. The same fate seems imminent for the survivor. The test tube is 10 mm in diameter. From Welsh 1984, Figure 3.23-24, with permission.

Lord Kelvin) in Glasgow a beautiful attempt to account for the existence of matter [Helmholtz 1867; Thomson 1867; Thomson 1883]. Helmholtz and Thomson were preoccupied with an incompressible three-dimensional fluid then imagined to pervade the entire Universe. This fluid, analogous to the "aether" of early electromagnetic field theory, was supposed to have inertia but to utterly lack viscosity, like liquid helium (see Box 9.D). The motions of any such fluid would naturally be organized by vortex lines.[3] According to this bold interpretation, each kind of atom is an ethereal vortex around a thread-like vortex core that closes in a ring—or in several such rings linked together. Each chemical element—each vortex atom—is distinguished by the topology of its aggregate of singular rings.

Though in its day it could be made to give an excellent quantitative accounting for some facts of atomic physics, the vortex atom model quietly went out of vogue thirty years later; and then atoms were found to have nuclei, so it stayed forgotten. But in a curious way, the most modern understanding of matter in terms of quarks, gluons, and so forth retains some of the flavor of the vortex atom theory: some particles (instantons and magnetic monopoles) are still supposed to consist of topological singularities in an otherwise featureless medium [de Vega 1978; Rebbi 1979; Bernstein and Phillips 1981; Nash and Sen 1983]. Particles in either context are viewed as defects in an otherwise smooth map from real space to the pertinent internal state space—in our case the circle of phase values. In excitable media we classify these defects mathematically by the taxonomy of periodic organizing centers and assume or demonstrate numerically that particle-like solutions to the equations of reaction and diffusion classify in the same way.

The same principles recur in a remarkable way on the grand scale of cosmology, in a theory of the origin of the Universe that leans heavily on the quantum-mechanical interpretation of nothingness—of the vacuum. This is the literature of "cosmic strings" [Kibble 1976ab; Vilenkin 1981, 1984, 1985]. According to one theory of the vacuum, the ground states of the vacuum are degenerate, lying anywhere along a ring of energetically equivalent phases. If you make a grand tour of inspection through regions devoid of matter, assaying the phase of the vacuum along a closed path through the Universe, your ship's log upon return may reveal that phase changed through some non-zero number of complete

---

[3] Vortex lines in fluid mechanics are geometrically, but not dynamically, analogous to our singular filaments. For example, each singular filament lies at the center of a pattern periodic in space and in time, while vortex flow has no periodicity in either sense; vortex lines can have any intensity while ours are quantized (being simply present or absent); and singular filaments can have any degree of local differential twist while vortex lines have none (being simply the center of a vortex with no marking wavefront).

Box 9.D: Vortex Rings in Air and Water

People see only what they are prepared to see.
—Ralph Waldo Emerson (1863)

At this point we have completely departed from our original preoccupation with periodic time. Vortex flow in fluids is not even periodic, as circulation is slower further from the vortex core. Nonetheless, principles encountered in answering one question often raise questions too intriguing to ignore in other areas.

The nineteenth-century vision of vortex atoms entailed linkage and knotting of vortex filaments in an imaginary fluid. That idea failed, but what about real fluids? We have all seen the equivalent of a scroll ring in the fluid we breathe: it is a smoke ring. Can smoke rings be linked? Knotted? Knots do seem possible in theoretical fluids [Moffatt 1969; Kida 1981]. How could such an exotic object be initiated? Would it be stable?

The answers may lie in the same geometrical principles exploited here with spatially and temporally rhythmic chemical reactions in mind. A sudden impulse perpendicular to the surface of a disk imposes a flow on the surrounding air, which inevitably circulates around the disk's edge: it creates a vortex ring. The ring moves, promptly getting clear of the disk. It moves because each segment of the vortex filament is the center of a cylindrical vortex that carries in its flow all the surrounding fluid, including remote parts of that same (curved) vortex filament. The whole ring thus transports itself in the direction of its own circulation through the ring [Kida 1981, 1982].

In the same way, an impulse applied along a twisted band (Figure 9.13, middle) would initiate a vortex ring along both linked edges; and an impulse everywhere perpendicular to the Seifert surface of a trefoil knot (Figure 9.13, bottom) would start a knotted vortex ring. How would such objects move (supposing the impulse-giving surface suddenly disintegrates so as not to obstruct continued circulation)? At each point along the filament, by calculus one can sum the components of fluid velocity contributed from the vortex surrounding each other segment of the filament. This may result in complicated tumbling and drift, but I think not in self-intersections: the filament should retain its integrity. The linked pair, for example, probably rotates in place about a 45-degree axle, while the trefoil propagates like a smoke ring.

Why have such exotic vortices not been observed in wind tunnels, mountain streams, smoke wreaths, or liquid helium? Just as in excitable media, one hears only of simple vortex rings. Because topologically nontrivial initial conditions are so unusual? Or because they are actually forbidden by some deeper principle, still overlooked? Or is it only because objects of unfamiliar symmetry escape recognition?

cycles en route. In this situation you confront a non-zero winding number again, but this time with cosmological implications. For the same reasons adduced in our laboratory three-dimensional media, this result would imply the presence within that loop of a one-dimensional filament along which the vacuum is not on the phase circle, hence not in the ground state, hence terribly energetic and thus possibly quite massive, per unit length, despite the absence of matter. These tubes of "false vacuum" are supposed to form closed rings, possibly interlinked and knotted, occasionally colliding and reconnecting much as do the singular filaments of excitable media. In this peculiar sense, a vial of the Belousov-Zhabotinsky reagent constitutes a tiny model of the Universe [Winfree 1986b].

Summary of Part III

Our pursuit of arrhythmias, paradoxically, has led to discovery of a new kind of rhythm, a mode of three-dimensional organization and timing in reactions that have the qualitative property of excitability. It turns out that such reactions are not so uncommon [Epstein et al. 1983], though they are still better known in living cells than in test tubes. Left to spontaneous processes, nothing much happens in an initially uniform excitable medium: it organizes itself in a featureless way, perpetually minimizing any concentration gradients. But when prodded by a big enough, spatially structured stimulus, it reveals alternative stable modes, organizing itself periodically in space and time. The result is a tiny structure woven of filaments not much thicker than living cells (about 150 microns in my laboratory at 37° C; 400 microns at 20° C in Müller's photographs [Müller et al. 1985; Figure 7.17 above]), rotating and pulsing and radiating waves of excitation that sequence the whole liquid's reactions in rhythmical patterns. The triggering stimulus is then no longer needed: the conjured structure is self-sustaining. This intricately periodic organizing center is evoked by stimuli initially designed to create arrhythmia in a spatially uniform oscillator!

The irony was already implicit in the cardiologist's term, arrhythmia. An "arrhythmia," we found in Part II, can be perfectly rhythmic: it just doesn't foreseeably return to the prior, normal rhythm. Rotating arrhythmia can be induced by a spatially graded stimulus. At the initial pivot of the rotor, the stimulus is singular: a disturbance of just the right strength applied at just the right time. Does "time" here necessarily mean only time in a cycle, as it did in Part I? We suggested a more general meaning by reinterpreting Figures 6.1 and 6.2 at the end of Chapter 7 (Figures 7.13 and 7.14) in Part II. The question may be posed more generally now.

If the job of the singular stimulus is to adjust simultaneously any two aspects of the system's chemical composition, the job might also be de-scribed as follows: wait until aspect A is right, then perturb selectively to bring aspect B right. If aspect A is ever right, it will be right at two times during the cycle of rise and fall of an action potential or a chemical wave. By increasing or decreasing B as required at either time, we have two cracks at phaselessness. This is the essential reason why we found two singularities in pacemaker tissues (Chapter 4). In this view, singularity is independent of the spontaneity of a system's oscillation; the critical *phase* for creating a pivot is really only a critical *time*, as we see in any one-shot excitable system such as heart muscle or the malonic acid re-agent. This interpretation of singularity as simultaneous adjustment of two quantities to critical values (see Appendix) corresponds to our inter-pretation of the singular filament as the intersection of two surfaces of uniform value.

In Part I we found that the critical annihilating stimulus inflicts ar-rhythmia on a biological oscillator. Thus it should also turn off an ar-rhythmia, such as an ectopic focus in diseased myocardium; clinical trials of this experiment were at least compatible with that idea. In Part II we found that it can also start oscillation in a new, spatially structured mode (the rotor). Applied in three dimensions, the critical annihilating stimulus might even be called a critical conjuring stimulus, creating in otherwise quiescent excitable media a spatially structured oscillator whose essence is the cyclic interplay of differently reacting regions.

These organizing centers are made out of three-dimensional versions of our familiar phase singularity, linked and woven together in beautiful symmetries. Their existence was not even suspected a decade ago. Though quarantined to the theorist's imagination during their first few years, they have recently escaped into the real world: chemists have synthesized the simplest ones and biologists are beginning to report objects of similar appearance and behavior in living tissues.

Each kind of organizing center is woven in its own distinctive way from fibers of phase singularity, as though from the funnels of chemical tornados. But what is moving here? Not matter: fluid flow, momentum, and viscosity are not involved. Rather, organizing centers are little chem-ical engines made of rotating parts, like most of the other rotating ma-chinery of our technology, except that the rotating parts are only patterns of chemical activity, like ghosts in the material substrate.

Like flames and crystals, these organizing centers provide laboratory examples of a chemical process that structures itself in a nontrivial dy-namic pattern. Unlike flames, crystals, and the diffusional instabilities already eagerly sought by theorists for thirty years, the organizing centers of excitable media seem indifferent to boundaries further away than a core diameter. In that respect they are more like independent particles. Also unlike the familiar objects of theory, these lack spontaneity: they arise only when triggered by some definite stimulus. But once started they have a life and stability of their own. They constitute the most tangible incarnations of phase singularities encountered up to now.

PART IV

Postlude

CHAPTER 10

What's Next?

The Past is but the beginning of a beginning, and all
that is and has been is but the twilight of the dawn.
—H. G. Wells
The Discovery of the Future (1901)

The kinds of material that lend themselves to the evolution of life also tend to develop rhythmic transformations, for example, internal clocks and even the life cycle itself. These transformations typically follow rules that, though unfamiliar, are nonetheless ubiquitous. They are unfamiliar largely because the behavior of biological and chemical oscillators has only been studied seriously since the early 1960s.

In 1964 one could find still unbound on the shelves of research libraries Zhabotinsky's seminal analysis of Belousov's reaction (in Russian) [Zhabotinsky 1964], Ghosh's discovery of oscillatory metabolism in yeast cells (in English) [Ghosh and Chance 1964], Wever's mathematical models of the human circadian clock (in German) [Wever 1962, 1963, 1964], Nagumo's observation of spiral waves in an electrochemically excitable system (in Japanese) [Suzuki et al. 1963], Moe's computer simulation of atrial fibrillation (in English) [Moe et al. 1964], and Perkel's strong-resetting data from pacemaker neurons (in English) [Perkel et al. 1964]. Twenty years later, what seemed isolated curiosities in 1964 seem in retrospect only the first points of contact with a connected, coherent body of phenomena organized by unfamiliar principles of timing and arrhythmia.

Printing such sentences near the end of a book gives the impression of ending a story. But science is like history: what we call the end will one day be called the beginning. What will happen next? Of course no one knows, if "next" means in five years or in ten. But most of the experimental demonstrations discussed in Parts I, II, and III were prefigured theoretically in some detail a few years before those dreams were tested in the laboratory. In fact, without theoretical support, experiments of the required delicacy might never have pinpointed a singularity, except as an occasional irreproducible accident. Does theory today prefigure at least vaguely the natural phenomena to be discovered tomorrow? Does a twenty-year avalanche of discoveries have any momentum; can it carry our imagination forward at least a few years to anticipate in outline the kinds of discovery most likely to present themselves next? Which research reports lying unbound on library shelves today will prove to be the seeds of books written a decade or two hence?

Let us try to probe a few years forward by simply extrapolating. Like snowslides, ideas and applications do tend to grow exponentially for a while with clear forward momentum. In fact the publication rate has doubled every three to four years, starting around 1964 and continuing still, in the subjects touching on biological and chemical oscillations. In order to extrapolate, let us get a running start by first recapitulating Parts I, II, and III:

We started with a view of our rotating planet, its time zones, and the internal clocks that mark us as having evolved on its surface. We found in our own physiology (using "our" in the most comprehensive possible sense, as the continuous heritage of life evolved on Earth) a convergence of time zones (isochrons) to a timeless Pole. We then found it again in shorter-period neural pacemakers. We found it again embedded in geographical context, this time in tissues made of clocks—and then found the clocks themselves dispensable: in excitable tissue the singularity became the rotating pivot of a spiral wave. We saw it not only in excitable tissue but in nonliving chemical media as well, no longer an abstraction about timing relations but now a visibly rotating source of waves. And there we first saw the singularity in its fullest development, as a tornado-like filament arching through three-dimensional space to close in a ring.

We began with daily time organization in a piece of rotating machinery, none of whose parts (mountains, oceans) has an intrinsic daily rhythm; but coupled together and pulling one another along in sequence, they constitute a clock. We passed through wide-ranging studies of chemical life forms evolved on the surface of that clock, finding many chemical oscillators. We found in each a center of rotation that must not be touched lest that clock's temporal organization be deranged. Embedded in physical space, we found that phase-scattered clocks typically reorganize in waves circulating about a center. We ended up with a piece of rotating chemistry, none of whose parts (droplets of reacting liquid) need have intrinsic rhythm; but coupled together and pulling one another along in sequence, they collectively constitute a clock.

At this stage we were dealing in snapshots depicting layered organization of reactions in space, no longer preoccupied with timing itself.

From a snapshot alone, who knows whether we are dealing with rhythms in time, too? Is time even necessary? Are there topological organizing centers of a purely spatial character? We will return to this question below.

The Geometric Outlook

The most conspicuous point, overall, is that biologists have found applications for a kind of inference used only infrequently outside pure mathematics and theoretical physics. This "kind of inference" is conducted by reasoning with symbols, particularly with pictures rather than with Greek letters, in the style of topologists (see the Introduction).

The sciences of life have never been admired for quantitative exactitude. I think one can proclaim loudly without fear of contradiction that the mathematics evolved to serve "the exact sciences" has only seldom been of critical use to biologists. But it cannot be said that living things are at heart sloppy, fuzzy, inexact, and unscientific. How does an oceanic salmon find its way home to spawn in the very rivulet it left in Oregon three years earlier? How is a meter-long sequence of billions of nucleotide base-pairs reversibly coiled without entanglement into a nucleus no more than a few thousand base-pairs in diameter? How does anyone memorize a vocabulary and rules of grammar well enough to transfer an epic novel from one brain to another? How does a gymnast calculate forces and rates for hundreds of muscles with millisecond precision while whirling from one maneuver to the next? How does a mere mortal perform a piano concerto—or compose one? Such miracles bespeak reproducible precision. But that precision is not the kind we know how to write equations about, not the kind we can measure to eight decimal places. It is a more flexible exactitude that evades quantifying, like the exactitude of a cell's plasma membrane dividing the universe into an inside and an outside with not even a virus-sized hole lost somewhere in all that convoluted expanse: topological exactitude, indifferent to quantitative details of shape, force, and time.

This is the kind of precision best described and reasoned about symbolically, rigorously, in the style of topologists. Just as physicists had to nurture a mathematics appropriate to the tasks they encountered, so must biologists. But the tasks are quite different in nature and call for mathematics of a different flavor. For biologists much of the beauty of topological reasoning is that it requires so few dubious assumptions; its few implications are correspondingly hard to avoid. Inference has power when it goes right to the heart of the matter without imposing on the biologist a tax for details that simply cannot be known at present and that in fact don't matter for the central point at stake. This is the kind of inference we used repeatedly in connection with circular patterns of timing, almost without regard for the underlying mechanisms. Our reasoning was based on qualitative features of laboratory experiments, such as winding numbers that could only be 0 or 1, a distinction that is hard to mistake and is unaffected by the inevitable quantitative differences in sensitivity and so on that make one individual organism different from another. Experimental conditions were specified quite flexibly. For example, in phase resetting by "a stimulus," the stimulus did not have to be an instantaneous impulse: any program of varying influences, no matter how prolonged and elaborate, fits the topological requirements if the ritual can be started at any arbitrary "old phase," elicits "even" resetting when played through to the end, and can be discontinued at any earlier stage. Our inferences seldom involved speculation about adaptive values, molecular mechanisms, or neural pathways. But they led us to ever sharper focus on experimental conditions in which something strange was guaranteed to happen: return of metamorphosing flies to the timeless condition of the newly fertilized egg, perpetual insomnia in mosquitoes, abrupt suspension of pacemaking in otherwise perfectly healthy and capable heart muscle, vortex centers of arrhythmia in electrically rhythmic tissue, chemically timeless rotors sequencing reactions around their perimeters, and chemical clocks made of shifting patterns of color topologically locked into three-dimensional organizing centers.

Discoveries to Come

In a certain sense the causes of these singularities are topological, not physiological or chemical: where the mechanisms are different, the phenomena still remain, incarnated in a different mechanism. Nonetheless, much—perhaps everything in some cases—remains to be understood about those underlying mechanisms. Obvious work for the near future includes elucidation of the particular nature and mechanism of arrhythmia in each context. It seems reasonable to expect that mechanisms of circadian rhythmicity—probably different mechanisms in each case—will be discovered in many organisms and cell types. Phase singularities in some will prove to derive from paralysis of each clock mechanism, in others from labile achievement of an unstable steady state in each cell, in others from a delicately balanced scattering of cellular rhythms all around the clock. What will that particular arrhythmia feel like to the first human suddenly emancipated from the cycle of sleep and wakefulness? What stimulus is capable of inducing such insomnia?

In spatially extended tissues, for example, at the center of a full-cycle phase gradient in a sheet of rhythmic fungus or in the broad leaf of a green plant, visible circadian clock singularities may be discovered. What becomes of such cells, perpetually exposed to all time zones at once? The singular arrhythmias of bioelectric tissues in particular invite understanding in terms of ion channels and spatial patterns of current flow. When the innervation of the human heart is better elucidated, someone may find three-dimensional patterns of nervous stimulation that accidentally instigate scroll rings. Or it may be found that the initial conditions for rotors or scroll rings arise from fluctuations of coronary microcirculation. Although detailed calculations must still be performed, it now seems more plausible that something awaits discovery here.

Topologically exotic scroll rings have up to now been observed only under the ideally controlled conditions of computer experiments. They will undoubtedly be synthesized in chemically excitable continua sooner than in cardiac muscle or brain tissue. Methods of initiation will probably involve topologically peculiar electrode configurations for microinjection of ions and currents, analogous to nerve and vasculature arborizations in living tissue. Once created, organizing centers may be expected to deform and move in sustained electric fields or chemical gradients. Their modes of decay—by shrinkage? by transmutation to simpler organizing centers?—will likely be revealed first by digital simulation and chemical experiments. Analogously linked and knotted vortices in fluid flow—elaborate smoke rings—may be discovered and observed to metamorphose even sooner.

We might also expect answers not only to the explicit challenges presented by singular arrhythmias, but also to an implicit question underlying the whole story. Throughout this book we clothed the winding number theorem in colorful diagrams. The fact that the costume fits is in itself worthy of wonder. Why is the human sense of hue so intimately associated with circularity that we can smoothly label points on a ring with unidirectionally varying hues and return to where we started without backtracking? The fact that color sensations span a simply connected space of two dimensions is easy to understand, given three spectrally different cone cell types; but why do we so naturally fiber that space on a circular base of "hue"? Does this have some neural basis in the millimeter-scale periodic organization of the visual cortex?

Singularities of Purely Spatial Patterns

Questions that seem already well formulated theoretically will be answered in the laboratory, given the ingenious resourcefulness of experimentalists and a few years' time; such, anyway, has been the pattern during the twenty years surveyed in this volume. What, then, about conceptual innovation? Dreams are a lot less predictable than programmable work, but dreams too have some direction, a momentum that can be extrapolated at least a little way. The direction is evident from the evolution of Part I to Part II to Part III. We began with clocks, discussing time independently of space. Then we introduced space, at first two-dimensionally. Meanwhile the medium's local capacity for spontaneous oscillation proved dispensable: the same phenomena arose in merely excitable tissue or reactions. Up to this point the underlying mathematics made no reference to complexity of mechanisms: any number of inter-dependent factors might be interacting to generate the oscillation or support excitability. However in turning to three-dimensional organizing centers, we quietly fell back to the case of just two dominant components in the chemical mechanism. If there are more, then singular filaments may behave quite differently, "fraying" at the ends, for example, instead of closing in rings. Might there be entirely different kinds of organizing centers, in no way related to rotors, spirals, and scrolls, but equally independent of boundaries, free, and particle-like? Must they be associated with phase, rotation, and cycles at all? Because this is a book about time, we were preoccupied with maps to the circle, maps that could be painted with hues from a color-wheel palette. There are quantities that are in no way associated with periodic time but take their values on a continuum with the connectivity of a circle; the supposition of one such underlay an era of excitement in the study of biological pattern formation (see Box 10.A). But there are also other kinds of space in which interesting quantities can take their values.

The fact that directions in two dimensions correspond to points on a circle (the horizon, a compass rose) gives rise in principle to a book on pattern formation analogous to this one, in which the words time, cycle, rhythm, and clock need never appear. The same singularities would arise, but with a wholly geographical meaning. Examples may eventually be discovered in the spatial-orientation centers of our brains.

A circle of compass directions or colors is not very different from the circle of phases. To take a really different case, remove the little direction arrow to consider just orientation of a line segment. A segment rotated 180 degrees is then the same as the original: the space of possible orientations differs topologically from the circle in that opposite points are equivalent. Something of this kind awaits elucidation in the visual centers of the primate brain, where columns of cells are committed to detection of line segments with preferred orientation. Could these preferences be arrayed two-dimensionally around phase singularities where electron transport is measurably different [Swindale 1982; Horton 1984; Horton and Hedley-White 1984]?

A more richly developed application is found elsewhere: in, for example, the microanatomy of fibrous materials like the cornea of the eye, the chromatin-packed nuclei of cells, the chitin shells of crabs, spiders,

Box 10.A: A Coat of Many Colors

Ten years ago a number of curious symmetries attracted the interest of theorists to the phenomena of limb regeneration. Regeneration is obviously a topic of fundamental importance to anyone who needs to know how animals grow, retain their form, and recover it after accidents. Regeneration of limbs in humans seems limited to fingertips, but we *did* grow full limbs once, and might again if medical scientists can figure out what inhibits that capacity in adults. Adult lobsters, insects, and salamanders, to mention only a few examples, regenerate lost appendages routinely.

This subject is still in the stage of becoming more opaque the more it is studied. But during the late 1970s one aspect of regeneration seemed (perhaps it was an illusion) suddenly to go translucent. This apparition was enhanced by the discovery that much formerly bewildering phenomenology could be succinctly summarized by topological mapping: a mapping, in fact, to the circle.

The symmetries that support this interpretation concern the number and handedness of extra organs or even whole limbs that often sprout from a zone where something has been removed or put back incorrectly. A tidy interpretation was put forward by Vernon French of the University of Edinburgh and Peter Bryant and Susan Bryant of the University of California at Irvine [Bryant et al. 1977; Bryant et al. 1981]. Called the clockface model, it associated a clock hour 1 through 12 with every point on the surface of a normal limb or organ. The number was not intended to represent time, but only position along a circle; 0 and 12 could not be

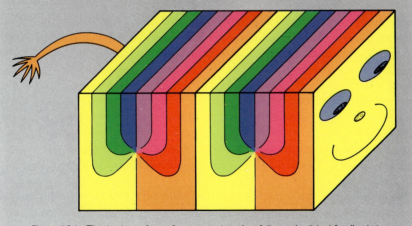

Figure 10.1: The body surface of any organism that follows the "clockface" rubric can be colored with hues from the cycle of saturated color sensations; limbs protrude under phase singularities.

distinguished. In other words, the covert states of normal tissue are taken to lie in a smooth ring-like state space of unknown nature. The number circle reaches around the circumference or perimeter of the normal region. A more convenient implementation of the same idea colors the skin in such a way that a full cycle of hues encircles the limb or organ capable of regeneration [Winfree 1984c]. The whole surface of an animal, then, is colored smoothly—except for a handful of left-right paired singularities (Figure 10.1).

A healed-over amputation or the grafting of a right limb onto the stump of a lost left limb (grotesque, but sometimes the best that can be done for an accident victim) is seen as a deletion of some part of the color field, followed by intercalation of a color patch of the same or opposite handedness. The result is anticipated as follows: The hues smooth out as tissues grow and mingle. If any singularities appear, a limb sprouts under each, being left- or right-handed according to the winding number [Glass 1977] around the singularity. This simple rule was so succinct and so successful in summarizing complex facts that it provoked experiments in dozens of laboratories to test the limits of its validity. As a result the whole matter is again shrouded in doubt and controversy. But the fact remains that developmental biologists conceived of a topological model of animal development and it served well for at least a few years. Other more sophisticated models await their turns.

and insects [Bouligand 1972; Gordon and Winfree 1978]. Such materials consist of spatially periodic arrangements of little fibers, arrangements inevitably organized around singularities akin to those seen in the timing of excitable reactions. Akin, but not identical. The difference in a two-dimensional context can be seen on the palm of your hand. Fingerprints are arrangements of oriented ridges about a millimeter apart, harboring singularities quite unlike the familiar radially converging isochrons [Penrose 1979]. There are two distinct types: one around which ridge lines loop and one where they meet tri-radially. A laboratory culture dish of skin cells typically harbors several such singularities along with the cells [Elsdale and Wasoff 1976].

We still have not departed very far from our monotonous routine of mapping things to circles. A greater departure comes with remembering that the materials mentioned are really three-dimensional, and that directions in space correspond not to a horizon circle but to the celestial sphere (with antipodes identified if opposite fiber orientations are indistinguishable). Maps from a spherical surface to the sphere of directions come in discrete variety, each associated with a winding number just like

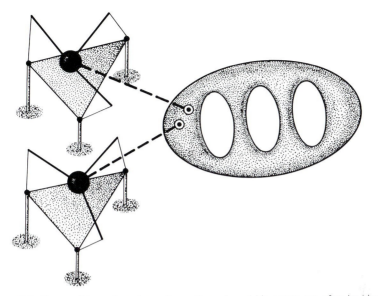

Figure 10.2: The possible configurations of a continuously variable system can often be identified with points in a continuum of appropriate dimension and connectedness. Here the configuration space of a three-legged gymnast is shown as the two-dimensional surface of a three-hole doughnut. Similar configurations map to adjacent points.

maps from a circular region to the circle of directions. Volumes in which the map type changes (as in our stimulus rectangle, containing both even- and odd-resetting maps) must have singularities. The geometry of these singularities in three-dimensional liquid crystals, as they are called, is a rich and beautiful subject with many biological applications.

Even so, we have scarcely ventured out of our familiar backyard. What about the sense of bodily orientation of a gymnast? Maybe this involves spaces and maps, somewhere in the brain, akin to those above. To take a much simpler example, consider a robot "gymnast" composed of a body grasping the corners of a triangular horse by three arms, each with an elbow. There are many possible configurations of this three-armed gymnast, and she can glide smoothly among them. Could they be mapped as points on a piece of graph paper? Yes, but it would have to be graph paper with no edges, shaped like a sphere with three tunnels smoothly bored through [Thurston and Weeks 1984] (Figure 10.2). This arcane example is not supposed to be obvious; but it makes the point that there are exotic topological surfaces associated with objects and behaviors that we usually describe haltingly in words, with no comprehensive insight. Geometric intuition provides an alternative path—perhaps the most natural path—to an understanding of complex and flexible "life-like" precision.

Many other quantities of interest that take their values in curiously connected spaces are familiar to physicists engaged in the study of quantum fields. One such space is associated with the electron, which turns out to be a different object after turning through 360 degrees but gets back to its original state after an additional rotation to a total of 720 degrees [Bernstein and Phillips 1981; Bolker 1973]. The flows of liquid helium are described in terms of the singularities of maps to spaces that are perfectly simple and orderly but almost impossible for the uninitiated to grasp [Maki 1978; Mermin 1979; Michel 1980]. Much the same development must be expected in the life sciences, as diligent investigators strive to quantify phenomena that up to now have resisted identification with such simple spaces as the line of real numbers. How are the various nuances of aroma classified? Aromas do vary continuously, just as colors do, but there are obviously more than a few variables involved and they may be connected in simple but subtle ways. The essential phenomena of odor perception may remain beyond our grasp until the topological nature of that space is grasped.

The simple topology of periodic time provided only meager illustration of this geometric approach. Yet it led us on a good chase and motivated more than a few surprising discoveries. What will happen on the next chase?

MULTIDIMENSIONAL APPENDIX

> Mathematicians are like Frenchmen: whatever you say
> they translate into their own language, and it means
> something entirely different.
>
> Attributed to Goethe

This Appendix attempts to connect our relatively model-independent reasoning about circles to the more sophisticated literature of mechanistic models that inevitably involve multidimensional geometrical structures.

About the Color Code

At this writing, no one knows how human color vision works, electro-physiologically. There are some empirical facts, however, well established in the case of uniform fields of color. One is that if you choose any three colors at random, no mixture of two of them in whatever proportions will duplicate the sensation of the third. By smoothly varying the proportions of all three, one obtains a two-dimensional continuum of distinct color sensations. (This is not a law of nature or a principle of physics, but just a fact about the manner in which our nervous system interprets the firing rates of our four kinds of light-sensitive cells: bees have one more kind of receptor, so they distinguish colors on a three-dimensional continuum; "color-blind" individuals have one less, so they have only a one-dimensional continuum of color sensations.) The practical consequence for our purposes here is that any non-self-intersecting closed curve on the two-dimensional continuum of normal human color sensations will pass through a cycle of colors, returning to the original color without ever repeating. That is all we need in order to color-code the phases of a cycle. It is not important that the colors be saturated, include purple, or enclose white or grey. However this is the choice made in the main text when a visual language for phase in a cycle is wanted: we use only the "hue" aspect of color, insisting upon full saturation.

A theorem frequently used in the text asserts, in these terms, that inside any such loop there must exist either a discontinuity in the color field or a color that is nowhere to be found on the loop. This is a mathematical theorem, not only a generalization from psychophysical experiments. It may be formulated more precisely as follows.

Consider a space that is locally like familiar Euclidean space (e.g. has no branch points or sudden changes of dimension), without holes or tunnels, in which you can travel from any place to any other without leaving the space (technically, a simply connected, path-connected manifold, M). For example, the three-dimensional volume inside a test tube is such a space. Next consider associating to each point of this space a color from the ring mentioned above, in such a way that points very close together in space have colors that are very close together on the ring (technically, a continuous map from M to the circle, S^1). Not all colors need be used, and the same color may be used for many points in space.

Now consider any closed path in this space and its coloring. Circumnavigate the closed path, noting the color at each point. Follow those colors along the ring-shaped palette. In particular notice how many times the color advances clockwise around the palette (the net number, subtracting any counterclockwise circuits). This number, the winding number of color around the closed path in real space, is necessarily an integer number, since we return to the same color. Perhaps more surprisingly, that number is necessarily *zero*. That is the theorem. It pertains only to continuous maps.

Thus if the number is obviously *not* zero, as in all applications in this book, then one or another of the assumptions must have been violated. The obvious choice is continuity: there is a defect in the map. Most of this book is about the geometry of such defects, in the context of biochemical and physiological phenomena.

This Appendix may be regarded as a dissenting opinion, the minority report of a truth squad following hard on the heels of the main text. It says that in every case the postulated maps to a ring should really be reinterpreted as maps to a palette of more dimensions—of color, of state of a perturbed circadian clock or electrophysiological membrane, of composition of a rhythmic reaction. As emphasized in the first papers to characterize a phase singularity experimentally [Winfree 1968, 1970ac, 1973d], the violated assumption is not really continuity but rather the very idea that the map was to a one-dimensional space, that phase alone adequately identifies the state of the underlying mechanism. The "defect" is then replaced by some color (or state) that was not originally contemplated only because it is not on the ring. Whatever colors are found on the ring, the "defect" color cannot be one of them. If all hues are represented around the ring, the "defect" must be hueless. But colors have more attributes than hue alone—for example, there are variously saturated colors of each hue. Because these colors are not on the ring, the "defect" can still be some color connected by continuous change to colors on the ring.

The version in the main text may be regarded as a procedure for finding phenomena that were unforeseen only as long as thought was limited to some ring of possible states: for example, the organizing center (a stable source of waves in a medium that would be quiescent if homogenized), the rotor (a periodic but nonoscillating mode of chemical reaction), the circadian clock's phase singularity (the labile arrhythmia of embryos, the insomnia of critically exposed mosquitoes). Perhaps certain arrhythmias that immediately precede sudden cardiac death will also turn out to involve such singularities.

Mechanisms

> Philosophy is written in this great book (by which I mean the Universe) which stands always open to our view, but it cannot be understood unless one first learns how to comprehend the language and interpret the symbols in which it is written; and its symbols are triangles, circles, and other geometric figures, without which it is not humanly possible to comprehend even one word of it; without these one wanders in a dark labyrinth.
>
> —Galileo Galilei (1623)

The main text of *this* book is built almost exclusively from circles, as is appropriate to a discussion of periodic timing. Our main finding was that circles are not adequate: a thoughtful consideration of any description based on circles reveals unavoidable situations "when time breaks down," that is, when no descriptive point can be found on the circle. By simplifying to a "circles" description, we isolated the points where that approximation fails; now we must backtrack a little to see what those points consist of in terms of the original, more correct and more familiar but more cumbersome descriptions.

We began above with circadian systems, which, as far as anyone has yet discovered, lack any important physical dimension of organization: circadian timing seems spatially uniform and is autonomous in the individual cell. It is therefore feasible to inquire into the dimensions of the circadian clock's internal state space without getting hopelessly confused by the additional dimensions of nonlocal variation. Various models (some as much as thirty years old: models tend to last a long time in this area) postulate one to four mutually interacting local quantities, thus making the internal state space one- to four-dimensional.

The existence of a timeless state of the circadian mechanism is implicit in many such models, being the necessary steady state of whatever re-actions are supposed to underlie timing. It is reasonably supposed that such a state would be difficult or impossible to reach, since all internal state variables would have to be simultaneously set to precise levels, and especially since the mechanism might linger only unstably near this state.

The phase singularity is a quite different concept of greater practical pertinence. As noted in Chapter 1, it was anticipated from three directions simultaneously:

(1) Any attracting limit cycle process (see below) involving any number of real-valued quantities such as chemical concentrations, regardless of complexity and the dimension of the state space, must have a *set* of arrhythmic states (generally not steady states). This phaseless set "has codimension 2 or less" [Winfree 1968; Winfree 1970c; Winfree 1974e; Guckenheimer 1975], i.e. some part of it can be reached by simultaneous adjustment of only two quantities.

(2) The notion of imposing a phase measure on observations of a rhythmic system, especially a population of cellular oscillators, implicitly entails the impossibility of assigning a phase under certain conditions that can be realized in a wide variety of two-parameter experiments [Winfree 1975a, 1976, 1979].

(3) Limited but high-quality phase-resetting data, regardless of any theory of mechanism or theory of measurement, seemed (in 1966) to reveal either a discontinuity of very nearly one cycle in the dependence of new phase on old phase or a continuous dependence in a topologically unfamiliar pattern (later dubbed "type 0," "strong," or "even" resetting). Choosing the continuous description implicitly focused attention on the topological anomaly that is now called a "phase singularity."

The coincidence of these three inferences was compelling enough to motivate some years of laborious experimentation, since repeated in many independent laboratories using different organisms. As a result of these collective efforts (Chapter 1), the reality of phaseless sets, phase singu-larities, time crystals, and so on became firmly established. Their phys-iological "meaning" is less clear, unless, contrary to convention, math-ematical concepts be accepted as valid "explanation" of biological phenomena. But that deficiency is in a way the most interesting aspect of these findings: because their prediction was in no way dependent on the mechanistic underpinnings of circadian physiology, the same prin-ciples might find application in other areas of physiology and biochem-istry.

These principles are not "mathematical," in the familiar sense of "mathematics" as "moving symbols around on paper" or "moving numbers around in computers." They are, rather, concepts about continuity that could be used in diverse contexts with sufficient rigor to precisely infer biological and chemical events that had not been observed. It seems unlikely that their utility has been exhausted; it may be worthwhile to allude to them explicitly, at least by citation, in this more conceptually oriented review of the main themes of this book. For a more formally mathematical review gleaned from the first draft of this book, see Strogatz 1984.

Phase

The circadian theme of Chapter 1 (dealt with more fully in Winfree 1980 and Winfree 1986a) provided an introduction and a way of dealing with spatially uniform patterns of timing. In the parallel development of descriptive models for neural pacemakers, the technique of space-clamping again ensures spatial uniformity, again allowing the investigator to focus his subtlety on the dimensionality of the local, internal state space. Thirty years ago it was already common for neurobiologists to think in the four-dimensional state space consisting of local membrane voltage and three factors of ionic conductivity. Many more channels confound modelers today.

Similarly, in efforts to unravel the dynamical complexity of chemical oscillators, three or four state variables are common in today's models, and some have a dozen or more.

But for some purposes, in all such contexts involving periodic dynamics, it is sufficient to restrict attention to the region of that multidimensional state space where the system spends most of its time. In periodic dynamics that region, like a circular subway tube, closely surrounds a closed loop embedded in the state space. For many purposes, then, the important part of the state space may be thought of in useful simplifying approximation as a one-dimensional ring of phase values.

But why simplify? In this book the main focus of analysis is on spatial organization of dynamics in physical space. Having additionally to contend with one dimension of physical space (as in plane waves, action potential propagation), or two such dimensions (as in rotors), or three (as in organizing centers), it was essential to restrict the local dynamical state space to the lowest practical dimension: one. Moreover, the discipline of formulating concepts in terms of the single quantity that is directly measurable in the laboratory (phase) pays off in terms of experimental design and anticipation of new phenomena. This is because "more realistic" multivariable dynamical models are usually hobbled by the unobservability of most of their interacting quantities. So the language of phase, coded into color, pervaded our analysis of spatial organization in progressively more and more physical dimensions.

But simplifications are approximate, and they may or may not seem too approximate, depending on the specific purpose envisioned and the taste of the individual thinker. For my own purpose—to anticipate and discover new phenomena, without pretense of formulating a definitive mathematical analysis—the phase approximation served well enough. But it would not be surprising if other discoveries await refinement of this approximation, as for example in the circadian (spatially zero-dimensional) application: the circadian clock's noncycling singular state lurked unnoticed in shadows offstage until the stage was set for its entry by adding necessary dimensions to the phase model, thinking about them topologically, and contriving laboratory experiments to detect the qualitatively unique implications [Winfree 1968].

What then may await discovery when, in the richer context of spatially structured dynamics, the simplified phase model is uncollapsed from a one-dimensional ring? More likely something than nothing. Part of it may have already nucleated in the form of a mathematical recipe for the initial conditions of organizing centers [Poston and Winfree 1987]. It seems appropriate to dwell briefly on the needed enhancements of realism, hinting at their mathematical structure wherever possible (mainly by citation).

As long as the internal state space was approximated by the single timing variable, phase, the notion of an "isochron" needed little explication: it is a locus of uniform internal state. But the purpose of this appendix is to contemplate a more realistically dimensioned state space. The notion of "isochron" now needs definition in terms of "state."

Isochrons in Physical Space

"Isochrons," as used in the main text, are loci of uniform timing traceable on a physical medium or on an abstract diagram describing a stimulus. Developmental biologists use the term for regions where pattern is simultaneously fixed, for example, in a growing peacock feather. Bird-watchers use the term to describe contours of seasonal reappearance of migratory birds on the globe. Physiologists use the term to describe loci of simultaneous activation in the heart muscle. As the name connotes, each isochron is a set of places or stimuli that result in something happening "at the same time." The "place" and "stimulus" interpretations were fused into one idea in Part II, where stimuli are graded across the physical medium.

Asymptotic vs. Immediate Isochrons

It should be noted that there are at least two senses of "at the same time" to distinguish with care. The original meaning, in the context of attracting limit cycles [Winfree 1967; Winfree 1970c et seq., summarized in Winfree 1980], is that the rhythms starting from any two distinct initial states on the same isochron eventually become indistinguishable. A more recent and often more useful meaning [Kawato and Suzuki 1978; Kawato 1981] makes no appeal to the limit of long time: two points are considered to be on the same isochron if a prescribed observable event happens next at the same time for both. This is closer to the cardiac physiologist's meaning.

State Space

These ideas are more powerfully formulated, not in the physical space of a rhythmic medium or in the descriptive coordinates of an arbitrary stimulus, but in the state space of the underlying mechanism. The instantaneous "state" of any local mechanism may be described by specifying the instantaneous values of the most important quantities whose mutual interaction constitutes the mechanism. For example, in the classical Hodgkin-Huxley description of a space-clamped membrane, there are four such quantities: (1) the membrane voltage, (2) a factor ranging between 0 and 1 that governs potassium channel conductivity, and (3) and (4) two more such factors governing sodium channel conductivity (see Figure 2.3). The state space is thus four-dimensional. All four quantities being real numbers, the four-space is Euclidean. In an appropriate simplification, the four quantities can be paired, resulting in a more approximate but more understandable projection onto Euclidean two-space [FitzHugh 1960, 1961; Scott 1975]. In any case, all preoccupation with state spaces and movement therein assumes that state varies continuously in time. Box 3.A questions whether this assumption will hold up in the final analysis; but in this appendix we sweep that issue under the rug.

The system moves through this space in a characteristic way, perhaps swiftly converging on a trajectory that leads in a closed loop (an attracting limit cycle) or in most of a loop, but arresting at a stagnation point short of reaching the beginning again (as in an excitable but not spontaneous system). In the still more drastic simplification used in the main text, that loop may be regarded as the only accessible part of the state space except during perturbations and except for isolated points caught inside a circle of phase.

Isochrons in State Space

Initiating the system at any point in its state space, one observes subsequent behavior. Those initial states may be classified by the timing of subsequent behavior, defining isochrons—but now in the state space rather than in the spaces used above. These different pictures are only distorted versions of one another. For example, the space of stimulus descriptions may be embedded in the state space because each stimulus leaves the mechanism to start at some definite state. Where this projection of the stimulus plane cuts through isochrons in the state space, they are "drawn on" the stimulus plane.

In the case of a mechanism with autonomous attracting limit-cycle behavior, far-reaching theorems can be rigorously proved about topological aspects of isochron arrangement—and thus about timing and its rearrangement by stimuli and so about its arrangement in physical space. Some of these theorems were conjectured from geometrical intuition and computational examples [Winfree 1974e], proved by Guckenheimer with appropriate qualifications [Guckenheimer 1975], and further discussed by Kawato and Suzuki 1978. They all concern the boundaries of the attractor basin, i.e. the set of states around the attracting cycle through which trajectories lead to that cycle. Trajectories necessarily cut through isochrons at unit rate everywhere. The limit cycle, like every other trajectory, punctures each isochron once, in order. Isochrons necessarily converge along the boundary of the attractor basin, each point of which is arbitrarily close to all isochrons. A part of this convergence locus must thread the limit cycle, in the sense that it punctures every two-dimensional surface bounded by the limit cycle. This boundary locus of "phase singularity" has codimension either 2 or 1 (i.e. of dimension 2 or 1 less than that of the state space, however great that may be). It has the larger dimension (codimension 1) when a steady state has a finite attractor basin around it, as in one version of the Hodgkin-Huxley equations [Best 1979] or when there are saddle-type steady states [Glass and Winfree 1984]. The phase singularity is the attractor basin's boundary, so it is also the boundary of the space outside the attractor basin that is called the "phaseless set" because it lacks isochrons. The phaseless set has the codimension of its bounding phase singularity, or less. For example, if an attracting steady state coexists and competes with the limit cycle for trajectories, its entire attractor basin is in the phaseless set, which thus has codimension 0.

The phase singularity is not to be confused with a steady state. It may border a phaseless set that contains attracting steady states. It may itself even contain nonattracting steady states (saddle points). Once in the phase singularity the system may approach such a steady state; but since the entire phase singularity is arbitrarily close to every isochron, it is clear that no such steady state can be stable against arbitrarily small fluctuations: they would perturb the system onto one or another isochron and thus launch it back toward the limit cycle.

In simple examples analytical solutions are available for phaseless sets and phase singularities and asymptotic isochrons [Winfree 1979; Winfree 1980]; they can be easily reworked for next-event isochrons. Unless the criterion for "an event" happens to coincide with one of the isochrons for asymptotic timing, the next-event isochron field turns out to be discontinuous along certain "cracks." Additional theorems for isochrons in this latter sense reveal the necessity of the discontinuities found [Kawato 1981]. The time is ripe for a definitive paper gathering what is now known of isochron geometry.

Isochrons without Spontaneous Oscillation

How can isochrons be pressed into service in connection with rhythmic activity in a medium whose local behavior is merely excitable, lacking a limit cycle? This has never been carried through in a way that lends itself to theorem-proving. However the idea finds employment in connection with organizing centers in excitable media [Winfree and Strogatz 1983a], and in connection with phase resetting in cardiac muscle [Honerkamp 1983; Strittmatter and Honerkamp 1984].

A notion of equivalent timing in this context may be obtained by extrapolation backward from a chosen observable event (e.g. threshold crossing or peak excursion from steady state) along a trajectory through the state in question to determine time since the event; this is the "negative latency." Or one might measure forward in time to an arbitrary criterion of closeness to steady state. Phase thus advances spontaneously at unit rate along trajectories up to 1, at which point the nonspontaneous excitable system is nearly settled into steady state. While the system still loiters near rest, we say the "phase" measure remains stuck at 1. Only a threshold-transgressing stimulus can move phase to 0, initiating the next cycle.

So much for the local, autonomous dynamics and its timing. We need next to encompass the role of a perturbing stimulus and the role of interaction with neighbors. First the stimulus.

Stimulus as Altered Flow Geometry

"Stimulus" means "an alteration of the dynamics": trajectories at each point in state space have differently altered components of velocity in some or all directions and thus generally lead in altered directions. If the system was initially on its limit cycle, it will generally move off the limit cycle during the stimulus. Consider a ring of initial conditions all along the limit cycle. The effect of a stimulus of fixed duration is to blow that ring off the limit cycle, creating a "windsock" of new intermediate states accessible by applying that stimulus for various durations. The moved ring is called the "shifted cycle" in Glass and Winfree 1984. As long as the stimulus is small enough, the shifted cycle scarcely differs from the limit cycle and therefore is still threaded by the phase singularity and therefore still punctures a complete set of isochrons with winding number 1. As coupling interval (alias initial phase) is varied through one full cycle (around the limit cycle), the new phase (the complement of latency modulo unit period) also varies through one full cycle. This is odd-type resetting, alias type 1 or "weak" resetting.

However, a larger stimulus (creating a longer windsock) may so distort the shifted cycle that it no longer encircles the phase singularity. The winding number is then zero and we have even resetting, alias type 0 or "strong" resetting.

If the phase singularity has codimension 1 rather than 0, then new phase can also change discontinuously along the shifted cycle, resulting in indeterminate winding number. But if the winding number exists, then only the two possibilities 1 and 0 can be achieved by prolonging exposure to a stimulus of given time-dependence in a state space of only two dimensions. Any other winding number would require the perturbed cycle to somewhere cross over itself, thus violating the supposition that the trajectory depends uniquely on the instantaneous stimulus and state. Other integer resetting types can occur in mathematical models involving three state variables (e.g. Winfree 1980, page 167) but have not yet been observed (or sought, perhaps) in nature. They are seldom observed even in complex models. For example, Wilson and Cowan 1972 and Enright 1980 show how neural systems composed of myriad cells lacking even resetting behavior and phase singularities may nonetheless en masse engender smooth limit-cycle oscillations that have even resetting and a phase singularity scarcely distinguishable from much simpler oscillators [Winfree 1975a, 1976, 1979; Enright and Winfree 1987]. Even time-delay oscillators (with an infinite-dimensional state space) have been found to behave quite prosaically in a pinwheel experiment [Johnsson and Karlsson 1971].

As a result, the simplest and most symmetric limit-cycle resetting model [Winfree 1970a; Winfree 1975b] has been widely adopted as a representative idealization [Guevara and Glass 1982; Hoppensteadt and Keener 1982; Taylor et al. 1982; van Meerwijk et al. 1984; Keener and Glass 1984; Hoppensteadt 1985]. This "radial isochron clock" was first used to fit (with remarkable, if meaningless, precision) data from about

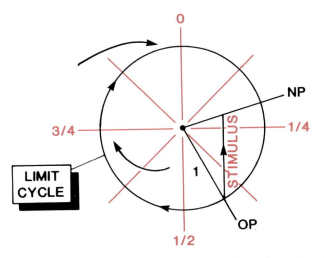

Figure A.1: Oscillation about a steady state here placed at the origin tends to unit amplitude, while angular velocity is strictly independent of amplitude. Isochrons (red) turn out to be equispaced radial lines. During stimulation the vertically portrayed variable is assumed to change much quicker than the horizontally portrayed variable: the change from one isochron to another is then simply described trigonometrically. Simple though it is, this picture captures the essentials of phase resetting.

3000 circadian rhythm experiments with flies [Winfree 1970ac]. It has some ungeneric features; for example, it would be more realistic to recognize that unless angular velocity is strictly independent of amplitude, isochrons wind infinitely often en route to the phase singularity rather than plunging in radially. But according to one philosophy, a model need not faithfully represent a region of behavior that defies experimental resolution anyway; thus it remains useful as a descriptive gimmick.

In this perfectly symmetric caricature of limit-cycle dynamics and its perturbation, phase resetting follows an analytically simple pattern:

The new phase is an angle NP related to the old phase angle OP by a triangle in Figure A.1: $\tan NP = \sin OP/(\cos OP + \text{STIMULUS})$.[1] Here NP and OP range from 0 to 2π in one complete cycle clockwise from north, and STIMULUS is some monotone function of increasing strength or duration of the perturbation. In neural context, positive stimulus means excitation (depolarization); inhibition (hyperpolarization) is negative. "Latency" is the complement of phase to one full cycle, taken modulo one cycle; "coupling interval" is the same as old phase, except perhaps for the units chosen.

Thus $NP = OP$ as long as STIMULUS $= 0$. The value STIMULUS $= \pm 1$ is critical: this positive stimulus (depolarization, excitation) applied at

[1] This basic relationship is often written in diverse ways (e.g. $\tan NP = (\text{STIMULUS} + \sin OP)/\cos OP$), depending on the direction of flow during the stimulus.

OP = 1/2 cycle makes NP undefined; applied negatively (hyperpolarization, inhibition) at OP = 0, it does the same. Taking phase 0 as upward threshold crossing before the action potential, and phase 1/2 as repolarization, this simple model provides a qualitative caricature of resetting behavior in the less symmetric limit cycles of electrophysiological pacemakers.

The same formula is reused in Chapters 6 and 7 as a caricature of Figures 4.1 to 4.5 and 4.9 to sketch the overall pattern of phase resetting in a tissue of neural pacemakers subjected to a spatially graded stimulus, as described next.

To describe the consequences of spatially graded phase resetting, we need to install a spatial pattern of stimulus intensity. This is done by replacing the constant STIMULUS by a diminishing function of distance from an electrode at the center of the square. The space constant for diminution is set to about 1 mm and the central stimulus magnitude is set to (+ or −) several times threshold, as in the experiment of Allessie et al. 1976 that induced paired rotors with such a (+) stimulus. A linear gradient of old phase (OP) or coupling interval is established from left to right. The stimulus and coupling interval are then used at each point in the square to compute a new phase (or latency) as above (Figures 6.1 and 6.2, 7.13 and 7.14).

The FitzHugh Model

In application to excitable membranes, this caricature has little to recommend it beyond simplicity. The next more realistic model, short of numerical engagement with the descriptive differential equations of multi-channel membrane models, is a distortion of this symmetric limit cycle that can also be derived from the Hodgkin-Huxley equations. This is the Bonhoeffer–van der Pol (as FitzHugh called it) or (more commonly) the FitzHugh-Nagumo model, often studied since FitzHugh [1960, 1961] highlighted its best features. Its phase-resetting contours in the pinwheel experiment resemble those seen in Figures 1.4, 6.2, and 7.14 (computed from the model of Figure A.1) but also exhibit distortions like pinwheel experiments seen throughout Chapter 4. Figure A.2 is a hand sketch emphasizing the asymmetries of flow relative to Figure A.1 and the consequent rearrangement of isochrons [Winfree 1979, Figures 5 and 6; Winfree 1980, Chapter 6, Box C; Scott 1979, Figure 8-5; van Meerwijk et al. 1984, Figure 8]. The most glaring contrast is in the bundling of about 1/2 cycle of isochrons near the trajectory that plays the role of "threshold," separating a full-blown action potential response from subthreshold perturbations. In Figure A.3 the consequences for resetting are shown in the case of a depolarizing stimulus by showing three successive displacements of the perturbed cycle:

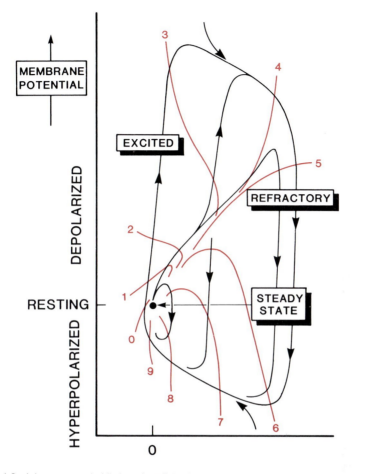

Figure A.2: A less symmetrical limit cycle, still involving only two variables, resembles a distorted version of Figure A.1. Isochrons (red) still cut the limit cycle at equal intervals—in fact all trajectories cross isochrons at equal intervals of time, by definition. All isochrons converge to the steady state.

(a) A slight offset upward (along the voltage axis, direction of depo-larizations) elicits smooth odd resetting, as the phase singularity is still inside the shifted cycle. As original phase (coupling interval) increases (clockwise around the shifted cycle), new phase also increases (latency decreases) smoothly though a full cycle.

(b) A moderate stimulus displaces the perturbed cycle further upward, where it no longer encloses the phaseless convergence of isochrons. Each encountered isochron is encountered an even number of times and some are never encountered. This is even resetting. On the arc of the perturbed cycle just prior to spontaneous depolarization, new phase is decreasing

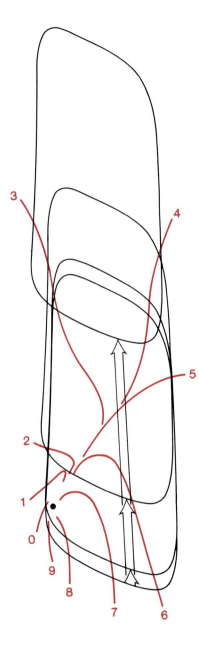

Figure A.3: Building on Figure A.2, "shifted cycles" are shown for stimuli of three sizes. The first and weakest elicits odd resetting; the second elicits even resetting with a near-discontinuity in the lower left, where a slight change of stimulus size or old phase (marked off along the unshifted cycle by red isochrons) results in displacement across many isochrons; the third and greatest stimulus elicits smooth even resetting.

as old phase increases; and the decrease is practically discontinuous where isochrons are close-spaced. Within a roughly 2-fold range of stimulus magnitude around the singularity, new phase varies quite abruptly with old phase near the critical phase. An electrophysiological model especially constructed from pacemaker-ball measurements may give a better accounting for these gaps than does the FitzHugh-Nagumo caricature [Clay et al. 1984]. The winding number (and curve "type") may be practically indeterminate in resetting curves assayed experimentally, which will inevitably show gaps where new phase appears to leap from one value to the next without intermediates.

(c) Larger stimuli (reached in cardiac pacemaker-ball experiments by van Meerwijk et al. 1984 and by Guevara 1984 but not by Scott 1979) still elicit even resetting. The isochron bundle has fanned out more substantially where it is now encountered; experimental resetting curves are correspondingly smoother.

It would be of interest to complete calculations for rhythmically driven atrial or ventricular muscle or Purkinje fiber, analogous to the calculations done from the Hodgkin-Huxley equations for Figures 4.1 to 4.5 and from a sinus node model for Figure 4.9. It would be even better to actually measure the offset of timing in a real preparation, as in Figure 4.10. At present no such information is available to see how nearly the qualitative features of these maps reflect actual adjustments of timing.[2]

Excitability without Oscillation

How does this picture metamorphose when parameters are slightly altered (and with them, the trajectories of Figure A.2) so that the slow trajectories near steady state actually flow into it and halt, as in Figure A.4, rather than passing by to go around again? The impact on qualitative behavior could scarcely be greater: the erstwhile pacemaker is now silent unless delicately triggered to escape the close environs of steady state and

---

[2] A first attempt was made by Chay and Lee [1984, 1985] incidental to their more sophisticated analysis of current-biased models for Purkinje fiber and ventricular muscle. Unfortunately for our interest in phase resetting, parameters in this study were carefully adjusted into a peculiar range where the phase singularity is drawn out along the borders of a large attractor basin surrounding steady state and where the basin's boundary is exquisitely sensitive to parameters (near a bifurcation point). The part of those publications concerned with phase resetting presents only contour maps (not the actual values), which contain enough inconsistencies to make interpretation difficult (results at phase 0 differ from results at identical phase 1, there seem to be large phase shifts at zero stimulus, individual computations often seem to be discontinuously out of line with surrounding computations). There remains much opportunity for clarifying the behavior both of electrophysiological model equations and of real tissue.

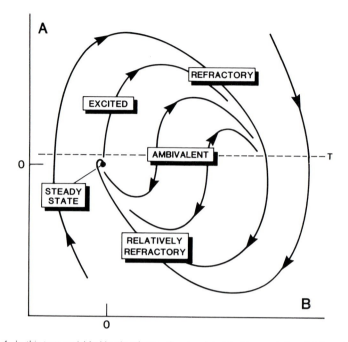

A

REFRACTORY

EXCITED

AMBIVALENT ———————— T

0

STEADY
STATE

RELATIVELY
REFRACTORY

B

0

Figure A.4: In this two-variable kinetic scheme closely related to Figures A.2 and A.3, trajectories near steady state become very slow and in fact fall into steady state. A slight displacement upward, though, launches the system onto another cycle of excitation.

undertake once more the long excursion that it formerly slipped into spontaneously again and again. So far as trajectories are concerned, the change is slight, indeed scarcely noticeable except very near to steady state. And what about isochrons? In the original definition concerned with asymptotic phase on a limit cycle, there are no more isochrons because there is no more limit cycle. Yet most of the "cycle" is still there: it has only been interrupted by a tiny gap near steady state. Except near steady state, the scarcely altered trajectories still flow across the old isochrons in equal increments of time everywhere. The old isochrons remain accurate landmarks for timing: they still provide a time-oriented coordinate for navigation in state space. A stimulus that perturbs the excited system during its return to steady state still displaces it from one isochron to another and so alters its negative latency (advances or retards its approach to steady state) by that much.

Moreover, the effect of a stimulus in terms of trajectories in state space remains unaltered, except that stimuli that start during the slow, near-steady-state portion of recovery from excitation will find the system initially somewhat nearer to steady state than it was in the spontaneously cycling version. In other words, whether the limit cycle is intact or is just barely broken, stimuli change the system's state in the same way, and the description of that change in terms of limit-cycle isochrons is still valid even when the limit cycle is interrupted.

Of course, this does not mean that behavior following the stimulus is the same: if the limit cycle is interrupted, then the phase-shifted system, considered as an isolated or spatially uniform system, will proceed only as far as that interruption and arrest there—until some further stimulus nudges it across the hiatus to the beginning of a new cycle. But in a medium of points all coupled to their neighbors by diffusion, that nudge is automatically provided, given the right spatial arrangement of initial states. That "right arrangement" is described in terms of isochrons by a convergence in rotary sequence, as in the pinwheel experiment. In Chapters 6 and 7 we saw that such initial conditions might lead to sustained circulation in a medium of points coupled to neighbors. The argument was that the zero isochron (the wavefront) would propagate forward, following the preestablished circular gradient of timing. We will now examine this supposition in terms of trajectories in state space.

Spatial Arrangements of Timing

How can a spatially graded stimulus, falling in the midst of a spatial gradient of timing, inaugurate circulating activity? This mechanism was first demonstrated conceptually in terms of an excitable wave-propagating reaction [Winfree 1973c] and computationally in terms of an excitable medium borrowed from neurobiology [Winfree 1974b; Winfree 1978]. This scheme is here briefly reviewed first for an oscillator like Figure A.1 (or Figure A.2), as in Figures 6.1 and 6.2, and then for a merely excitable system like Figure A.4 as in Figures 7.13 and 7.14.

Suppose first a rectangular medium as in Figure 6.1 with an east-west gradient of phase. A stimulus gradient will be added now, not in the radially symmetric way of Figure 6.1, but more simply as in the basic pinwheel experiment: stimulus duration varies from 0 at the south edge to a maximum (which need not be brief) at the north edge. (A similar argument can be made for stimulus strength; in fact the character of the stimulus can also change as duration is prolonged, without violating the topological essentials of this kind of argument.) Stimulating the entire rectangle simultaneously, we displace each point in it from its original state on the limit cycle to someplace along a trajectory that continues to lead away from the limit cycle as long as the stimulus persists. In Figure A.5, the oscillator's y variable (perhaps membrane potential) is assumed to increase at fixed rate during the stimulus.

Thus the rectangular area of the medium is mapped into a (flattened) cylindrical region of state space, as shaded. This is the pinwheel experiment, viewed not in physical space but in the internal state space. Instead of showing the state (or just the phase, a function of state) at each point

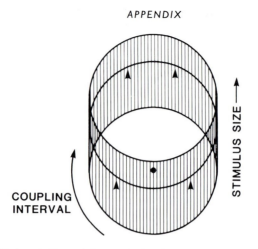

Figure A.5: Much as in Figure A.3, perturbed cycles are shown on the model of Figure A.1. A cylindrical continuum of stimulus sizes is shown by shading.

in physical space by contour lines, we are showing the place at each point of state space by contour lines of latitude and longitude.

The stimulus being finished, each oscillator continues its normal kinetics from whatever state it has reached during the perturbed kinetics. It may seem a gratuitous assumption that there is no aftereffect of exposure to the stimulus in this illustration. But in fact there is no loss of generality. To say that normal kinetics recovers only slowly under standard conditions after the stimulus is removed is to say there are additional variables of state that gradually relax back to normal values: i.e. the flow has additional dimensions. In that enlarged state space, the multidimensional flow is again an instantaneous function of the stimulus: in the absence of any stimulus, flow is strictly normal. The argument presented here in a two-variable state space would proceed identically in a space of more dimensions.

Let us suppose that the normal flow very swiftly leads back to the limit cycle. As is intuitively obvious (and formalized by the non-retraction theorem), flow must diverge from some point in the interior of the limit cycle. This presents no problem to independent points of the rectangle if they share no more than a common history with neighboring points. But if the rectangle is really a medium, so that neighboring points interact (e.g. by molecular diffusion or spread of electrical potential) then between any two nearby points is another point of intermediate state. As points diverge, concentration gradients or potential gradients increase without limit until further divergence is restrained, maintaining continuity.[3] Thus

[3] Coupling through molecular diffusion is represented by the Laplacian operator in the pertinent partial differential equations. Such a term may be present in the rate equation of

part of the medium remains tautly stretched like a membrane over a hoop while all surrounding points rotate along the limit cycle. This central rotating zone of steep gradients is the "rotor" for this oscillating, but inexcitable, local kinetics. Its period is nearly the same as the period of local autonomous oscillation. With varying degrees of convincingness, it has been much studied analytically in this perfectly symmetric approximation [Cohen et al. 1978; Hagan 1982; Koga 1982; Krinsky and Malomed 1983; Welsh 1984; Welsh and Gomatam 1984; Kuramoto 1984; Malomed and Rudenko 1986].

Is Reentry Stable in the Absence of a Hole?

Is this pattern of activity stable? The only way it can collapse to spatial uniformity is for all points to gather together on the limit cycle. And that can occur only if points presently along some arc of the limit cycle are prevailed upon to pull free of it inward, crossing the interior to the far side. If electrotonic or diffusive interactions are strong enough, i.e. if the physical medium is small enough and gradients are steep enough, then this will happen. However if the physical edge of the medium is remote from the center of the rotor then this action only translates the rotor physically across the medium. What would give direction to this putative migration and what would direct it to the medium's boundary? Until the entire medium pulls free from one side of the limit cycle (until the rotor migrates to the boundary) the rotor cannot cease to exist.

A similar picture may be drawn using the less symmetric excitable oscillatory kinetics of Figure A.2. In this case also the mechanisms of continuity act most forcefully where flow is most divergent, between mid-cycle and steady state. But now the limit cycle also comes close to steady state. Most of the cycle's period is accounted for by the slowness of flow during this perihelion; the remainder of the orbit, during excitation, transpires relatively swiftly. It is usual to find that this slow interval is abbreviated by the effect of neighbor coupling: points already launched into accelerating excitation develop large concentration/potential gradients by which tardy neighbors are encouraged away from their dalliance with steady state. Thus the rotor's period is shorter than the local oscillation would be in the absence of a rotor. And the rotor's pivot is no longer the local steady state: it is in the middle of the limit cycle. Semi-analytical approximations for the period, wavelength, stability, etc. have been obtained from mathematical models of the FitzHugh-Nagumo type [Mikhailov and Krinsky 1983; Krinsky and Malomed 1983].

---

each species involved, if they are all equally mobile. In the case of electric potential diffusion in cell membranes, the other state variables typically are relatively immobile, being the configurations of membrane-bound channel proteins that do not touch one another.

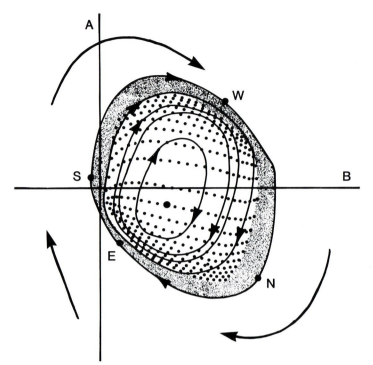

Figure A.6: In this snapshot, a square grid of 3,721 cells rotates on the state space of an excitable reaction between quantities A and B. Cells remain within a hoop defined by the trajectory of uniform excitation from steady state and back to steady state. Steady state is at the coordinate origin. The south, west, north, and east corners of the square are indicated. Five concentric circles on the square appear as closed rings here, each indicating the sequence of states traversed at the common period of rotation by all cells on that circle. Cells near the larger dot just below threshold remain at fixed composition at the center of the vortex. All but the 600 cells explicitly shown are far from the vortex center and are crowded onto the uniform excitation hoop; they are here indicated collectively by shading. Adapted from computations in Winfree 1978.

What if Figure A.2 mutates to Figure A.4, an excitable kinetics without spontaneous oscillation, rupturing the limit cycle but leaving most of the flow negligibly altered? In this modification, the steady-state point from which local flow diverges is again not central to a surrounding circulation but is a threshold locus near the endpoint of all trajectories. Nothing of great generality can be asserted mathematically, as there are all degrees of "excitability," including such weak degrees that even the propagation of a pulse is not sustained (see Box 7.C). However, in the dozen or so excitable media that have been explored numerically for interest in their ability to propagate waves and to support rotors, the picture remains much the same as above:

In the absence of neighbor coupling, the entire rectangle folds up like a torn scarf being stuffed into a pocket, and collapses into steady state along the trajectories of local kinetics [Winfree 1974b]. But given a continuity mechanism, numerical studies show persistent rotation of a disk of medium whose image in state space is tautly stretched over a cycle (Figure A.6). As in the previous case, the pivot is not the local steady state but is roughly the middle of the loop of trajectories that constitutes the normal course of one excitation.

Physically, an intuitive explanation is that most points initially approach steady state along the loop trajectory, but those that were initially just below threshold attempt a more direct approach. Being coupled to neighbors above threshold, they fail: they are pulled across and then pull their neighbors across. As in the previous case, time lingering near steady state is much abbreviated by this coupling. In this case that time is infinite in the uncoupled local kinetics; but with coupling, every point is caught up in a cycle whose period is roughly the duration of excitation and relaxation back to excitability. Some of these notions can be made quantitative through analytical approximations [Mikhailov and Krinsky 1983]. Numerical investigations of such rotors are reported by Pertsov et al. [1984].

This scene rather resembles an Escher drawing of a waterfall in a castle moat, maintaining circulation. If the rectangular medium had a hole in the middle (the castle), this picture would not surprise us. If it has no hole (if there is no castle, and the moat is really a lake), then the waterfall must taper toward the center in the manner of a Whitney cusp [Winfree 1973c]. The continued circulation is then a little more surprising, but not unreasonable, and it turns out to be stable in diverse numerical experiments. I believe it constitutes a reasonable interpretation of reentrant excitation in chemically excitable media such as the malonic acid reagent, in comparable biochemical media (layers of slime mold cells or layers of neurons subject to potassium-activated potassium release), and in blocks of cardiac muscle capable of smooth propagation in any direction.

The language of isochrons used throughout the text provided a simple language for describing this process in terms of observable timing, unburdened by the extra conceptual apparatus of a state space whose coordinates are often only "observable" by elaborate inference from indirect measurements. A proper theory would, of course, carry the burden. But I can do little more than sketch the outline of such a model, and appeal to numerical simulations [Winfree 1974b] and chemical experiments [Winfree 1985; Figure 7.8] as evidence for its reality. The advantage of completing it explicitly would, of course, be to clarify the limitations of this scheme. There are many mechanisms whereby reentry can be induced and it may be that only a minor proportion of them are usefully conceived in these terms. Without a quantitative theory, one cannot know.

Chemical Reentry in Three Dimensions

In the text the simple language of resetting an oscillator without neighbors was retained beyond Chapter 4, the last province of its native country, as a mere expediency for getting through the no man's land of Chapters 5–7 about coupled media without spontaneous local oscillation. Weary from evading hostile natives with mumbled assurances that we were guaranteed safe passage between two legitimate kingdoms, we arrived safely (?) in Chapter 8. Here we were again able to adopt the appropriate local dialect, in this case to carry on commerce about spatial patterns of phase. There was no mention of mechanisms, internal state spaces, or coupling in Chapters 8 and 9: the language of that country concerns only arrangements of timing in real physical space.

Nonetheless, processes encountered in Chapters 5–7 underlie those arrangements: organizing centers are composed of rotors connected by propagating waves, both of which depend on local excitability and coupling by diffusion. There is a multidimensional state space behind the scenes of any phase description, interpreting phase into chemical concentrations. While it is not yet prudent to attempt detailed examination of their rates of change, we might at least assure ourselves that any snapshot of an organizing center has some consistent interpretation in terms of those real-valued state variables.

This is done in Winfree 1985 and its elaboration Winfree and Strogatz 1983abc. Walking alternately on the two legs of phase description (as in the main text) and state-space description (as in this appendix), a picture of organizing centers is built up that is compatible with the requirements of physical chemistry and with the observation of local periodicity. But in retrospect the notions of differential geometry employed in those papers (e.g. writhing and twisting) seem superfluous. The more concise analysis given in Chapter 9 uses an index theorem about defects in the otherwise smooth map from real space to the circle of phases. Here we prove that those maps are compatible with real concentration gradients by direct construction, using a trick of complex-valued algebraic geometry.

Seeds for Organizing Centers

The mathematics is unfamiliar but the concept is simple. To create an organizing center, specify an initial arrangement of chemical concentrations that resembles the final, stable object closely enough: hopefully its stability properties will refine the initial conditions. A good beginning would be to make sure that the critical level of one reactant ("A" = A*) intersects that of the other ("B" = B*) (supposing there are only two of importance) along appropriately linked and knotted closed rings, and that concentrations vary smoothly everywhere. This is almost as easily done as said, thanks to a trick invented by Tim Poston of the Crump Institute for Medical Engineering at UCLA.

Poston's idea is to describe the concentration field analytically first in *four* dimensions, where it is symbolically convenient, then in a three-dimensional subspace of appropriate topology, and finally in the Euclidean three-space of a physical medium. First the local composition (A,B) is renamed $\underline{Z}$ and thought of as a complex number $\underline{Z} = (A - A^*) + i(B - B^*)$. Thus (A*, B*), the composition of the singular filament, is called $\underline{Z} = \underline{0}$. Then $\underline{Z}$ is described as a polynomial function of two other complex numbers, $\underline{X}$ and $\underline{Y}$. The space of ($\underline{X}$, $\underline{Y}$) thus has four real dimensions. The locus of roots of $\underline{Z} = \underline{0}$ is the locus of singular filaments. If $\underline{Z}$ is a factorable polynomial then this is the locus (a two-dimensional surface in the four-space) along which any one of the factors has its real and imaginary parts both zero. This is a beginning, but to see where it leads, we must first get back to real three-dimensional space.

The job begins by restricting attention to a three-sphere embedded in the four-space: the locus of points whose coordinates, squared, sum to a constant. This particular restriction to a lower-dimensional subset is chosen for symmetry and for the nice feature that it intersects all those two-dimensional surfaces of $\underline{Z} = \underline{0}$ along rings (loops), each linked through the others to constitute a single organizing center.[4] To each factorable polynomial corresponds a unique organizing center from among the list of those allowed by the exclusion principle (sum $L_{ij} = 0$ for each i).

For example, $\underline{Z} = \underline{X}$ is an infinitely long scroll ring, for all local purposes indistinguishable from a straight untwisted scroll.

$\underline{Z} = \underline{Y}$ represents an untwisted scroll ring of unit radius.

$\underline{Z} = \underline{X}^2 + \underline{Y}^3$ turns out to represent a single appropriately twisted ring tied in a 3:2 knot drawn on the surface of a unit torus. This initiates the continuous version of the organizing center computed in discrete approximation from discrete initial conditions as shown in Figure 9.13 (bottom). With a minus sign instead of the plus sign it would be the mirror-image knot.

---

[4] The term "organizing center" is here used in the chemical sense, but in this mathematical context we run the risk of confusion with the catastrophe theory sense of the term [Thom 1972; Thompson and Hunt 1977; Zeeman 1977]. In that sense there is also an organizing center *point* of our polynomial at the center ($\underline{X} = \underline{Y} = \underline{0}$) of the three-sphere, which dictates the topological features of all the surrounding maps.

$\underline{Z} = \underline{XY}$ represents a pair of linked twisted rings, as also does $\underline{Z} = (\underline{X} - \underline{Y})(\underline{X} + \underline{Y})$; the distinction between the two will come out in a moment. With a minus sign, each inverts through a mirror.

Next, we project this three-dimensional picture into real three-space. One projects a map of the globe (a two-sphere) stereographically to a flat map (Euclidean two-space) by placing a transparent globe on a tabletop, South Pole touching, and shining a light in all directions from the North Pole to the table surface. Each ray projects from the North Pole, to the place where it punctures the spherical surface, to the "shadow" of that place on the plane. The same is done algebraically in the three-dimensional case. In this way the North Pole is spread to infinity, along with any nearby concentration gradients. Thus in the flat projection, regions distant from the South Pole become uniform: exactly as needed for a local organizing center floating in an ocean of unstructured excitable medium that promptly settles to the steady state. Still better, circles and spheres in the spherical three-space map in this way to circles and spheres in real three-space. Some of them may be enormously expanded, depending on the exact orientation of the polynomial relative to the stereo projection. In this way $\underline{XY}$ becomes a twisted scroll ring threaded by a straight axial singularity (the linking ring): the organizing center that was studied analytically in linear approximation by Gomatam 1982 and computed from reaction-diffusion equations by Panfilov and Winfree 1985. The polynomial $(\underline{X} - \underline{Y})(\underline{X} + \underline{Y})$ becomes a pair of finite linked rings, the continuous analogue of the discretized version computed by Winfree et al. 1985.

For a specific example, take the untwisted scroll ring $\underline{Z} = \underline{Y}$, for which $\underline{Y} = \underline{0}$ is the locus of rotor pivots. How does it look in real physical $[u,v,w]$ space? First, what is the projection, algebraically, from the four-space $[Re(\underline{Y}), Im(\underline{Y}), Re(\underline{X}), Im(\underline{X})]$ to the three-space $[u,v,w]$? First we pick out our three-dimensional sphere-surface subspace as:

$$Re(\underline{Y})^2 + Im(\underline{Y})^2 + Re(\underline{X})^2 + Im(\underline{X})^2 = 1.$$

Then we place the North Pole at $Re(\underline{Y}) = 1$ and the South Pole on the table at $Re(\underline{Y}) = -1$; then by analogy to the projection of a globe onto the tabletop plane (scaling by 2), we write:

$$u = Re(\underline{X})/(1 - Re(\underline{Y}))$$
$$v = Im(\underline{X})/(1 - Re(\underline{Y}))$$
$$w = Im(\underline{Y})/(1 - Re(\underline{Y}))$$

So

$$u^2 + v^2 + w^2 = (1 + Re(\underline{Y}))/(1 - Re(\underline{Y})).$$

That is to say, each parallel $(Re(\underline{Y}) = $ constant) of the three-sphere in four-space projects to a two-sphere in real physical three-space; for tidiness, temporarily abbreviate this sum of squares as R^2. Then in our complex polynomial equation wherever we have:

$$
\begin{array}{ll}
\text{Re}(\underline{Y}), \text{ we may write} & (R^2 - 1)/(R^2 + 1) \\
\text{Im}(\underline{Y}) & 2w/(R^2 + 1) \\
\text{Re}(\underline{X}) & 2u/(R^2 + 1) \\
\text{Im}(\underline{X}) & 2v/(R^2 + 1)
\end{array}
$$

To evaluate concentrations $(A - A^*)$ and $(B - B^*)$ at physical point $[u,v,w]$, we obtain from these formulae the values of $\underline{X}$ and $\underline{Y}$, and use them to compute the appropriate complex polynomial $\underline{Z}$. Then, given suitable scaling coefficients a and b, we simply say:

$$
\begin{aligned}
A - A^* &= a \text{ Re}(\underline{Z}) \\
B - B^* &= b \text{ Im}(\underline{Z})
\end{aligned}
$$

If instead we want to know where in $[u,v,w]$ space the chemical composition has a certain value, we write

$$(A - A^*)/a + i(B - B^*)/b = \text{the complex polynomial,}$$

substituting for the real and imaginary components of $\underline{X}$ and $\underline{Y}$ their equivalents in terms of u, v, and w as above. So for example if we only want the singular locus, only for polynomial $\underline{Z} = \underline{Y}$, the answer is:

$$0 + i0 = \text{Re}(\underline{Y}) + i\text{Im}(\underline{Y}) = (R^2 - 1)/(R^2 + 1) + i2w/(R^2 + 1)$$

so

$$
\begin{aligned}
0 &= (R^2 - 1) = u^2 + v^2 + w^2 - 1 \\
i0 &= i2w \qquad \text{so } w = 0
\end{aligned}
$$

so

$$1 = u^2 + v^2, \text{ a horizontal circular singular ring.}$$

Appropriate polynomials, processed by complex number arithmetic in a personal computer, thus provide the numerical initial conditions for reaction-diffusion computations. Such seeds of three organizing centers are shown in stereo Figures A.7abc, as parametrized and plotted by Poston using an IBM PC/AT with an IBM Professional Graphics Display. They have all been used for rough calculations of chemical organizing centers in a PC/XT clone (the COMPAQ DESKPRO) with enough RAM to accommodate two chemical concentrations in each of $40 \times 40 \times 40$ cells. The computations show that the time-evolution of these snapshots, guided by local excitability and coupling to neighbors by diffusion, does indeed fulfill the expectation of Chapters 5–7. More refined calculations require the resources of a supercomputer: a project currently in hand.

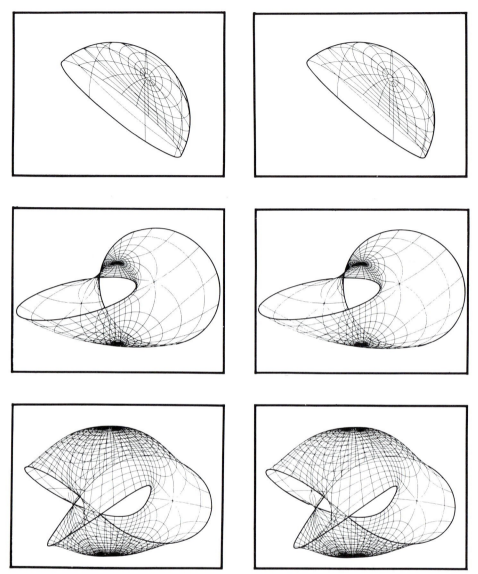

Figure A.7: Stereo snapshots of the wavefront A > 0, B = B*, of three seeds of organizing centers, as computed from complex-valued polynomials: (top) untwisted scroll ring; (middle) linked twisted rings; (bottom) trefoil knot. The left image of each pair is intended for the right eye. To arrange this, look at a pencil point held in front of your eyes, about halfway to the book. Instead of two images, you see four. Adjust the center two into register as a single 3-D image. From Poston and Winfree 1987.

Conclusion of Appendix

Imagine a spatially extended dynamical system (perhaps a pulsing jelly-fish, an oscillating chemical solution, an electric motor) that sponta-neously tends to some characteristic rhythmic behavior. Suppose some kind of perturbation upsets this rhythm, but it usually recovers, with some offset of phase that depends on the timing and manner of pertur-bation. If even resetting is observed, then, as we have repeatedly seen, the non-retraction theorem (or less general mathematical apparatus) guarantees that some fainter application of the same stimulus at a critical moment must result in completely indeterminate timing of that recovery (or no recovery at all). This phase singularity might be on the border of an alternative mode of activity (a different rhythm or even quiescence) or it might merely be a labile state from which arbitrarily slight fluctua-tions may at any time initiate return to normally organized rhythmicity.

A vertebrate heart is apparently such a system. But the phaseless states implicit in the mere observation of even resetting are probably of little physiological interest. Why? Because stimuli that inflict stable phase re-setting must do so by resetting the sinoatrial pacemaker—but the ar-rhythmias of greatest clinical concern are ventricular. In mammalian hearts the atrial muscle mass and the ventricular muscle mass are elec-trically isolated but for a single point of contact in the atrioventricular node. If rhythmic activity in ventricular muscle were autonomous (per-haps due to activity somehow reverberating in the mass of individually nonspontaneous fibers) then topological theorems might have deep sig-nificance for disruptions of cardiac timing. But the ventricular muscle mass, considered in isolation from its pacemaker node, is not sponta-neously rhythmic. It is driven from the remote sinoatrial pacemaker, which is largely protected from retrograde stimuli originating in the ven-tricles.

In this situation the non-retraction theorem applied to the whole heart describes only a kind of arrhythmia—probably a fleeting and unimportant kind—associated with stimuli that cause the sinus node to stumble briefly. The persistent and pernicious arrhythmias that evolve from direct stim-ulation of ventricular muscle may eventually reset the sinoatrial pace-maker, but that is incidental. The essential idea here is not that reentry arises by phase-shifting the sinoatrial pacemaker (though the title of Winfree 1982a gave that impression to some) but rather:

(1) That the immediate effect of a stimulus is practically the same in the state space of spontaneous (pacemaker) and nonspontaneous excit-able membrane;

(2) That the lasting effect of an appropriately timed stimulus gradient

in spontaneous (pacemaker) tissue would be rotation even in the absence of neighbor coupling;

(3) That numerical demonstration from model equations can be replaced (under certain assumptions) by topological proof in this case, because the multidimensional state space can be handled in terms of the single phase variable and its isochrons;

(4) That the effect of neighbor coupling is to maintain continuity and thus to stabilize and accelerate the rotor so formed;

(5) That in nonspontaneous media the state-space effect of graded stimulation and of neighbor coupling are essentially the same as in locally spontaneous pacemakers, though their description no longer lends itself to the same proofs because in the absence of a limit cycle the local dynamics has no isochrons and no single-variable summary.

Chapters 5–7 presented an analysis of reentrant, circulating excitation based on the analogy between rhythmically driven excitable media and spontaneously rhythmic local pacemakers. It is important to remember that the little mathematics involved was pertinent to the pacemakers, not to the analogy. The analogy is offered not as a rigorous tautology but as a simpler way of understanding complex empirical phenomena. A simpler way is needed if we are ever to comprehend rhythmic spatial organization and its breakdown in three-dimensional media. Fibrillation in particular is perplexing. I first noticed the word "fibrillation" seven years ago and I still don't know what it means. The accounts I have found in diverse literatures do not resolve my confusion. The effort sketched in this book doesn't either. It is a mystery. Somebody should solve it.

GLOSSARY

Words set in SMALL CAPITALS are defined elsewhere in the Glossary

ACTION POTENTIAL   The unit pulse of electrical conductivity by which an excitable membrane fires in response to a stimulus. Pacemaker membranes fire spontaneously even in the absence of a discrete stimulus.

ANGINA PECTORIS   Temporary ISCHEMIA in the heart muscle, sometimes painful, often caused by spasms of the coronary arteries.

ARRHYTHMIA   An abnormality in a normally rhythmic system's behavior: the PHASELESS condition of a LIMIT-CYCLE OSCILLATOR, e.g. a CIRCADIAN clock. In this biomathematical usage and as used by cardiologists, "arrhythmia" does not necessarily connote nonperiodic behavior: it only means return to the usual rhythm is not predictable. In cardiology some prefer the term DYSRHYTHMIA.

ATRIA (sing. ATRIUM)   The smaller "priming" chambers (left and right) of the heart, with relatively thin walls. (Formerly "auricles.")

ATTRACTING CYCLE   Short for attracting LIMIT CYCLE, a closed-loop trajectory in state space to which nearby trajectories converge from all sides throughout a volume called the basin of attraction.

AXON, GIANT SQUID   One of a pair of long nerves, about 1 mm in diameter, which mediate the squid's escape reflex and the insights of electrophysiologists.

BLACK HOLE   As used here, a HUELESS region in a LATENCY DIAGRAM that is colored everywhere else to indicate timing of a rhythm. Huelessness indicates that no specific timing can be associated with that range of stimuli: timing may be random or an ARRHYTHMIA may be induced, including as one possibility creation of a high-frequency ROTATING WAVE or, as another, quiescence.

CARDIOVERSION   Application of an electric pulse to the heart for purposes of terminating an ARRHYTHMIA and, it is hoped, restoring the normal sinus rhythm.

CHEMICAL WAVE   A propagated disturbance of chemical reaction rate—e.g. the burning of a fuse. Usually connotes a characteristic speed, as opposed to mere DIFFUSION (the passive spread of ink in water, for example) and as opposed to locally PERIODIC activity that does not really spread but merely appears to propagate due to fortuitous GRADIENTS of timing. A propagating change from one stable state to another, as in the burning fuse or a chain of dominoes, should be distinguished from the kind of chemical wave described in this book: a propagating pulse-like excursion from a stable state back to the same state. See also WAVE.

CIRCADIAN   Literally, "roughly daily," the adjective applied to physiological rhythms that continue in the absence of external cues, maintain a constant PERIOD near 24 hours nearly independent of constant temperature, and respond to visible light. (Organisms evolved on Titan would presumably have a circadian period of 16 hours or thereabouts, if the feeble sunlight makes any difference to them.)

CODIMENSION   The complement of the dimensionality of an object. For example, a point has codimension 2 in a plane, but only codimension 1 in a line. A line has codimension 2 in three-dimensional space. A sphere has codimension 0 in three-dimensional space. A PHASE SINGULARITY has codimension either 1 or 2 in whatever space it inhabits.

COLOR CODE   Metaphor for quantities defined on a circle, such as compass direction or PHASE in a cycle. Points, called hues, on the defining circle are colored in a way that smoothly runs once through the artist's/psychophysicist's saturated COLOR WHEEL without backtracking.

COLOR WHEEL   The sequence of adjacent perceived hues at maximum saturation: red orange yellow green blue violet purple red orange etc.

CONDUCTION BLOCK   The event in which a propagated ACTION POTENTIAL in excitable membrane encounters a region through which it cannot (or does not) pass in that direction at that moment, resulting in extinction of that arc of the WAVE. "Conduction block" is not an observation so much as the absence of an observation that might have been expected by naive extrapolation. Extinction of a wave at a boundary of the medium is usually not called "conduction block."

CONTINUITY   The relation between two quantities such that a sufficiently small change in one of them corresponds to no more than a small change in the other. See also DISCONTINUITY.

COPHASE   The complement of the PHASE of a rhythm (or clock, OSCILLATOR, etc.). The LATENCY minus any number of complete periods contained within it, measured in units of one PERIOD. If a rhythm is at PHASE 0.3 (fraction of a cycle elapsed), then its cophase is 0.7 (fraction of a cycle remaining).

COSMIC STRING   An essentially one-dimensional locus in the interstellar vacuum along which the vacuum cannot remain in a low-energy state: possibly a PHASE SINGULARITY in the field of vacuum states.

COUPLING   The connection of one dynamical system to another, e.g. of a tiny volume element to neighboring volume elements in a medium of EXCITABLE material.

COUPLING INTERVAL   No relation to "COUPLING," this is jargon for the time when a stimulus arrives, measured from some prior event, usually an ACTION POTENTIAL. The term derives from clinical cardiology, in which two beats (a "couple") were often seen to occur together at a characteristic interval, the "coupling interval." Normalized to the PERIOD, this is the same as OLD PHASE.

CRITICAL STIMULUS   A rightly timed STIMULUS of the right size applied to a PERIODIC system that responds with EVEN PHASE RESETTING to a larger stimulus of the same kind. The result is indeterminate timing after the stimulus: possibly random return to the prior mode, possibly a new mode (e.g. if an ORGANIZING CENTER is created), possibly quiescence. The "right size" criterion is commonly left uninvestigated and undemonstrated.

DEFIBRILLATION   The effect of any procedure for restoring normal SYNCHRONY in cardiac muscle, for example, by violent electroshock or by perfusion with potassium salts to depolarize uniformly.

DIASTOLE (dye-ass'-toe-lee)   The interval of relaxation that follows SYSTOLE (contraction of the ventricles) in the heartbeat.

DIFFUSION   Movement of something from a region of abundance into a region of scarcity in a way that (ceteris paribus) would bring each region closer to the average abundance in neighboring regions. Adjacent chemical reactions are coupled by molecular diffusion (and by other processes, such as radiation, in situations more complex than examples treated in this book); adjacent patches of nerve membrane are coupled by diffusion of electric potential (i.e. the GRADIENT of an electric current). Spontaneous eruption of concentration patterns due to *unequal* diffusion of two or more mutually reacting quantities is called "diffusional instability."

DISCONTINUITY   A value of one quantity at which a second, dependent quantity abruptly adopts a new value, strictly without any continuous transition, no matter how rapid. See also CONTINUITY.

DYSRHYTHMIA   Alternative jargon for ARRHYTHMIA, lacking the unwanted connotation of *non*rhythmicity.

ECTOPIC FOCUS   A point-like region of more rapid spontaneous activity in an EXCITABLE MEDIUM; a spurious pacemaker.

ELECTROCARDIOGRAM   A record of tiny electrical potential differences felt on the surface of the body due to electrical activity in the heart muscle. The signal at a point on the body surface is a sum of oriented dipole fields emanating from the boundary surfaces (epicardium, endocardium, INFARCT fringes) of the excitable tissues. A positive contribution is given by every bit of heart surface along which a wavefront of depolarization is approaching the body surface electrode.

ENTRAINMENT   The locking of one rhythm to another with N cycles of the one matching M cycles of the other (N, M being integers). The commonest mode of entrainment is of course 1:1. See SYNCHRONIZATION.

EVEN RESETTING   A pattern of PHASE RESETTING in which NEW PHASE (or COPHASE or LATENCY-modulo-PERIOD) varies continuously with no *net* change as OLD PHASE (or COUPLING INTERVAL) is scanned forward through one cycle.

EXCITABILITY   The ability of a system to respond to a sufficient STIMULUS by some large excursion from its prior relative quiescence and then to return to that prior state. There are many degrees and kinds of excitability, a subtlety of some importance that is swept under the rug in this book.

EXCITABLE MEDIUM   An excitable system distributed in space with COUPLING between adjacent pieces so that excitation can propagate as a WAVE.

FIBRILLATION   A complicated and poorly understood high-frequency mode of electrical activity that can be triggered abruptly in normal heart muscle and just as abruptly terminated. It can be triggered more readily in tissue that is appropriately "depressed" (slower depolarization and propagation, briefer excitation and refractoriness), especially in a spatially complicated way, especially by spatially graded stimuli. Though often described as utterly chaotic, fibrillation often seems to have a strong pattern. There may be different kinds of fibrillation. See also VENTRICULAR FIBRILLATION.

FLUTTER   A very regular high-frequency mode of activity in heart muscle (a TACHYCARDIA) that often precedes FIBRILLATION. Flutter probably involves a single ROTATING WAVE of electrical activation. Can occur in ATRIUM or in VENTRICLE (and is immediately fatal in the latter case).

GRADIENT   A region of continuous change in some local quantity from place to place.

HODGKIN-HUXLEY EQUATIONS   The "grandfather" equations of electrophysiology, the prototype for exact description of the dynamics of voltage and of specific ion conductivities in cell membranes.

HUELESS   The equivalent of PHASELESS in our color metaphor. A hueless point may have color, but not any of the colors used to encode the standard circle of PHASE.

INCOHERENCE   Failure of mutual ENTRAINMENT among the parts of a system.

INFARCT   A dead volume of heart muscle, killed by severe and prolonged ISCHEMIA.

ISCHEMIA   Deficiency of blood (oxygen, glucose) flow to an organ, especially used in connection with heart muscle.

ISOCHROME   An ISOCHRON colored according to our COLOR CODE for PHASE.

ISOCHRON   The curve connecting points that represent same timing of a rhythm: simultaneous activation, or same NEW PHASE or same LATENCY-modulo-PERIOD after a STIMULUS, etc. For example, cotidal contours, ISOCHRONAL CONTOURS, isochrons of epicardial activation, the isochronal lines of seasonal bird migration.

ISOCHRONAL CONTOUR (on an ISOCHRONAL MAP)   Term used by physiologists for the locus of simultaneous activation of heart muscle: a wavefront. This is one kind of ISOCHRON.

LATENCY   The time from one event to another, here used for time from the end of a STIMULUS (return to standard conditions) until the next ACTION POTENTIAL (or the second, or the third, etc.). In this book we usually measure latency relative to an extrapolation of the prior rhythm, as percentage of a normal cycle PERIOD, neglecting any whole number of repeats of the period.

LATENCY DIAGRAM   The contour map (often color-coded) of NEW PHASE or LATENCY as a function of STIMULUS magnitude and timing.

LATENT PHASE   Same as NEW PHASE.

LIMIT CYCLE   The rhythmic cycle of transformations that a system tends to return to after almost any small perturbation. Sometimes called ATTRACTING CYCLE.

LINKED SINGULARITIES THEOREM   Given a collection of mutually linked rings of PHASE SINGULARITY with ring i linked L_{ij} times through ring j, a twist L_{ij} must be locked into each SINGULARITY ring equal and opposite to the algebraic sum of that ring's other L_{ij}'s. In other words CONTINUITY of PHASE everywhere except on the SINGULAR-FILAMENT rings requires that the sum of all L_{ij} must be 0 for each ring.

LINKING NUMBER   The signed integer that describes the linking of one ring through another (same, not negative, if the two rings are taken in opposite order).

MALONIC ACID REAGENT   The Belousov oscillating reaction, as modified by Zhabotinsky (replacing citrate by malonate) and by Winfree (suppressing spontaneous oscillation while EXCITABILITY remains).

MEANDER   The spontaneous wandering of the ROTOR, the pivot of a ROTATING WAVE. Sometimes observed if the kinetics is only marginally excitable or very "stiff" (time scale of excitation very short relative to time scale of recovery); but otherwise, rotation is commonly rigid.

MEMBRANE POTENTIAL   The voltage difference (commonly 100 mv or less in living cells) across the semipermeable membrane bounding a volume of relatively immobile polyelectrolytes in an ionic solution.

MYOCARDIUM   The (involuntary) muscle of the heart.

NEW PHASE   The PHASE of a PERIODIC system at any integer multiple of normal PERIODS after a STIMULUS ends. (This was the original "Winfree" definition: occasionally used by others since to mean *one* period after, and occasionally "after a stimulus *begins*.") Same as LATENT PHASE.

NON-RETRACTION THEOREM   No manifold can be continuously retracted to its boundary (the soap-film-popping theorem).

ODD RESETTING   A pattern of PHASE RESETTING in which NEW PHASE varies continuously with a net change of one cycle forward as OLD PHASE or COUPLING INTERVAL is scanned forward through one cycle (or LATENCY-modulo-PERIOD undergoes a net decrease of one period as OLD PHASE or COUPLING INTERVAL increases through one period). This is always the result with a small enough STIMULUS, if behavior is continuous at all.

OLD PHASE   The PHASE of a PERIODIC system at the moment a STIMULUS program begins. See also COUPLING INTERVAL.

ORGANIZING CENTER   In two dimensions, the ROTOR. In three dimensions, the rotor drawn out into a SINGULAR FILAMENT (closed in a ring unless boundaries intervene). A group of mutually linked rings may be called a single organizing center. The term "organizing center" is used in a related sense in developmental biology to signify a piece of tissue without which the normal pattern never develops; and in a sense related to that in catastrophe theory about the singularities of gradient flows [Thom 1972; Thompson and Hunt 1977; Zeeman 1977]. In this book we are preoccupied instead with the PHASE SINGULARITIES of *non*gradient flows and their rather different organizing centers.

OSCILLATOR   Any system capable of stable rhythmicity even in the absence of external rhythmic cues.

PERIOD   The interval between two consecutive repeats of a rhythmically recurring event.

PERIODIC   Characterized by a regular repeat interval.

PHASE   A measure of the instantaneous state of a PERIODIC process: the fraction of a normal cycle already elapsed. See also COPHASE. Beware confusion with a multitude of alternative meanings, some remote and some only subtly different, e.g.: oil, water, and ice are three phases in physical chemistry; alternating current may be delivered by 1-phase, 2-phase, polyphase connections; the phase of a rhythm (the whole rhythm, not an instantaneous state concept) is its phase-difference relative to another; LATENT PHASE of an OSCILLATOR when it is not on its normal periodic cycle is the phase it approaches after many normal periods.

PHASELESS   The condition of a (potentially) PERIODIC system in which no component of that rhythm remains, e.g. if quiescent or if a different rhythm has taken over. There is a set of phaseless states of codimension 2 or less linking any ATTRACTING LIMIT CYCLE in a state space equivalent to Euclidean space. The boundary of the phaseless set is the PHASE SINGULARITY. The phaseless set is the complement of the LIMIT CYCLE's attractor basin.

PHASE-RESETTING CURVE   A plot of the NEW PHASE reached in a PERIODIC process as a function of OLD PHASE when a standard stimulus is given. (Or LATENCY-modulo-PERIOD vs. COUPLING INTERVAL.) Both coordinates being circular, this curve is really a closed ring, best plotted on a TORUS. The only curves observed in laboratory experiments up to now link the torus once forward (odd, type 1, weak) or do not (even, type 0, strong).

PHASE-RESETTING SURFACE   A stack of PHASE-RESETTING CURVES in order of STIMULUS magnitude (duration, size, quality), making a doubly PERIODIC surface in this three-dimensional coordinate space.

PHASE SHIFT   The advance (positive) or delay (negative) of next (or n'th) firing after a STIMULUS, expressed as fraction of a cycle, neglecting any complete cycles. The NEW PHASE minus the OLD PHASE, plus or minus an appropriate number of normal PERIODS.

PHASE SINGULARITY   A point (in real three-dimensional space, in an imaginary state space, or in the coordinate space describing a STIMULUS) near which the full cycle of ISOCHRONS crowd together. It is possible to have a line of singularities, as in the SINGULAR FILAMENT of an ORGANIZING CENTER or along the border of a BLACK HOLE. This SINGULARITY is not closely related to the singularity of catastrophe theory, which concerns itself with potential functions and nonperiodic dynamics [Thom 1972; Thompson and Hunt 1977; Zeeman 1977].

PINWHEEL EXPERIMENT   A way of systematically arranging a PHASE-RESETTING experiment in space, with an OLD PHASE (or COUPLING INTERVAL) gradient transverse to a STIMULUS-size GRADIENT.

PREMATURE BEAT   An electrical event in the heart that occurs before the normal expected time of next firing; often caused by an ECTOPIC FOCUS.

PURKINJE FIBER   A specialized muscle fiber, more like a nerve fiber, which rapidly conveys excitation to the inner surface of the VENTRICLE. (In some nonhuman species, Purkinje fibers may penetrate deeply through the muscle, too.)

REENTRY   The cardiologist's term for return of a WAVE of excitation to a place passed before, with reexcitation of that place. Reentry may be PERIODIC if the same path is always followed. Distinguish one-dimensional (ring) reentry (the kind contemplated almost exclusively until the 1970s) from two-dimensional (vortex) reentry (a possibly more realistic kind contemplated since then) from three-dimensional (ORGANIZING CENTER) reentry (still largely imaginary).

REFRACTORY   Unresponsive to attempted excitation by a chosen standard STIMULUS. An excitable system is refractory for a time after excitation; a WAVE has a unique forward direction because it has a refractory wake. Not the same as "unresponsive in terms of PHASE RESETTING," since a delaying STIMULUS will nonetheless fail to evoke an ACTION POTENTIAL.

RESETTING CURVE   See PHASE-RESETTING CURVE.

RESETTING SURFACE   See PHASE-RESETTING SURFACE.

ROTATING WAVE   A WAVE that propagates along a succession of fibers in a closed (one-dimensional) path; or that propagates around the perimeter of a (two-dimensional) hole, or around an area of functional CONDUCTION BLOCK, or (which might be the same thing) around a ROTOR, radiating away as a more-or-less symmetric spiral; or that propagates by curling (three-dimensionally) around the wavefront's edge like a scroll of paper.

ROTOR   The chemical or electrochemical source structure immediately surrounding the pivot of a ROTATING WAVE in two or three dimensions.

SCREW SURFACE   A smooth surface with a helical boundary around a central axis of PHASE SINGULARITY. Common in PHASE-RESETTING SURFACES.

SCROLL RING   The scroll-like WAVE surrounding a SINGULAR FILAMENT in an EXCITABLE MEDIUM, when the filament closes in a ring. Scroll rings are distinguishable according to integer TWIST. They may be knotted or mutually linked.

SEIFERT SURFACE   A smooth surface fitted (by Herbert Seifert's procedure) to a boundary composed of closed rings.

SINGULAR FILAMENT   A one-dimensional continuum of PHASE SINGULARITIES in three-dimensional space, like a vortex line.

SINGULARITY   Something special. In dynamical systems theory, it usually means an isolated state at which nothing is changing (equilibrium or steady state); or in catastrophe theory, the critial combination of parameter values at which a system's behavior or morphology acquires a new feature [Thom 1972; Thompson and Hunt 1977; Zeeman 1977]. In this book we use only PHASE SINGULARITY.

SINGULAR STATE   Vague term referring to the PHASELESS condition of an organism after appropriate STIMULUS. Not to be confused (if possible) with the equilibrium or steady state of a dynamical system or with the SINGULARITIES of catastrophe theory [Thom 1972; Thompson and Hunt 1977; Zeeman 1977].

SINOATRIAL NODE   A collective of tens of thousands of electrically coupled cellular OSCILLATORS near the superior vena cava in the mammalian right ATRIUM: the normal initiator of each heartbeat. The sinoatrial node is capable of PHASE RESETTING and may also support reentrant WAVE vortices under appropriate conditions.

SPIRAL WAVE   The WAVE of excitation emitted by a ROTOR, conducted in an essentially one-dimensional way along a line tangent to the rim of the rotor. Usually an involute spiral.

STIMULUS   An alteration of a system's environment in a way that has some effect on the system's time course. For our purposes the stimulus need not be brief or even constant; it may be a prolonged program of diverse influences.

SUDDEN CARDIAC DEATH   Unexpected abrupt failure of blood circulation to the brain for reasons to be sought within the heart. In most cases the cause is VENTRICULAR FIBRILLATION.

SYNCHRONIZATION Same as ENTRAINMENT for the purposes of this book; elsewhere it sometimes connotes ENTRAINMENT in 1:1 mode with essentially no PHASE difference (between two rhythms so similar that this is an unambiguous description).

SYNCOPE (sin'-cup-pee) Fainting due to temporary loss of blood circulation to the brain because of a transient ARRHYTHMIA of the heartbeat.

TACHYCARDIA (tacky-kar'-dee-ah) Abnormally rapid resting pulse (by definition > 100 beats per minute).

THEOREMS: NON-RETRACTION (see above); LINKED SINGULARITIES (see above).

TIME CRYSTAL A three-dimensional plot of the RESETTING SURFACE, PERIODIC along both OLD PHASE (OR COUPLING INTERVAL) and NEW PHASE (or LATENCY) axes. The third axis represents stimulus size.

TOPOLOGY Topology (meaning point-set topology so far as this book is concerned) is a collection of thoughts about continuous sets of adjacent points (e.g. surfaces) and continuous mappings of one onto the other. For examples, see THEOREM.

TORUS A doughnut, bagel, inner tube. Sometimes used to mean the whole volume, sometimes only its surface.

TWIST The integer LINKING NUMBER of a closed curve drawn on a SCROLL WAVE adjacent to the SINGULAR FILAMENT. To be distinguished from the non-integer "twist," Tw.

VENTRICLES The larger and more muscular chambers (right, and especially the left) of the heart.

VENTRICULAR FIBRILLATION "The mechanism of death in the majority of patients who die suddenly of nontraumatic causes" [Gettes 1984]: a sudden, pervasive, fine-grained loss of normal coherence in the rhythmical coordination of the muscle of the VENTRICLES. See also FIBRILLATION.

VORTEX ATOMS A hypothetical accounting of over a century ago for atoms as concatenated vortex rings in an aether-like fluid.

VULNERABLE PHASE A moment in the normal cycle of a biological clock when a STIMULUS of the right kind and size produces indeterminate PHASE RESETTING or lasting ARRHYTHMIA. Stimuli of different kinds have different vulnerable phases. In the case of the heartbeat, the vulnerable phase is a window of less than a tenth of a second in each cycle when a small local depolarization can trigger transition to FIBRILLATION. If the stimulus is too big or too small or wrongly timed, it won't work. There is an atrial vulnerable phase and, later, a ventricular vulnerable phase.

WAVE   For our purposes, a propagated disturbance that moves as a pulse at a standard speed (usually diminished close behind a forerunner wave or where the wavefront is sharply curved) through connected parts of a uniform medium. For example, see CHEMICAL WAVE.

WINDING NUMBER   The algebraic net integer number of full cycles of PHASE traversed along a closed loop path along which phase is defined and varies continuously.

REFERENCES

Abakumov, A. S., Gulyayev, A. A., Gulyayev, G. A. (1970) Propagation of Excitation in an Active Medium (in Russian). Biofizika 15(6), 1074-1080.

Abildgaard, C. P. (1775) Tentamina Electrica in Animalibus Instituta. Societatis Medicae Havniensis Collectanea 2, 157-161.

Adgey, A.A.J., Devlin, J. E., Webb, S. W., Mulholland, H. C. (1982) Initiation of Ventricular Fibrillation Outside Hospital in Patients with Acute Ischaemic Heart Disease. Br. Heart. J. 47, 55-61.

Agladze, K. I., Gorelova, H. A., Zurabishvily, G. G., Michaelov, A. C., Panfilov, A. B., Tsiganov, M. A. (1983) Control of Autowave Sources (in Russian). USSR Academy of Sciences at Puschino. Offprint.

Alessi, R., Nusynowitz, M., Abildskov, J. A., Moe, G. K. (1958) Nonuniform Distribution of Vagal Effects on the Atrial Refractory Period. Am. J. Physiol. 194(2), 406-410.

Allessie, M. A., Bonke, F.I.M. (1978) Direct Demonstration of Sinus Node Reentry in the Rabbit Heart. Circ. Res. 44, 557-568.

Allessie, M. A., Bonke, F.I.M., Schopman, F.J.G. (1973) Circus Movement in Rabbit Atrial Muscle as a Mechanism of Tachycardia. Circ. Res. 33, 54-62.

Allessie, M. A., Bonke, F.I.M., Schopman, F.J.G. (1976) Circus Movement in Rabbit Atrial Muscle as a Mechanism of Tachycardia: II. The Role of Nonuniform Recovery of Excitability. Circ. Res. 39, 168-177.

Allessie, M. A., Bonke, F.I.M., Schopman, F.J.G. (1977) Circus Movement in Rabbit Atrial Muscle as a Mechanism of Tachycardia: III. The "Leading Circle" Concept. Circ. Res. 41, 9-18.

Allessie, M. A., Lammers, W., Smeets, J., Bonke, F., Hollen, J. (1982) Total Mapping of Atrial Excitation During Acetylcholine-Induced Atrial Flutter and Fibrillation in the Isolated Canine Heart. In: Atrial Fibrillation, ed. H. E. Kulbertus, S. B. Olsen, M. Schlepper. Molndal, Sweden: Lindgran and Soner.

Allessie, M. A., Lammers, W.J.E.P., Bonke, F.I.M., Hollen, J. (1984) Intra-Atrial Reentry as a Mechanism for Atrial Flutter Induced by Acetylcholine and Rapid Pacing in the Dog. Circ. 70(1), 123-135.

Allessie, M. A., Lammers, W.J.E.P., Bonke, F.I.M., Hollen, J. (1985) Experimental Evaluation of Moe's Multiple Wavelet Hypothesis of

Atrial Fibrillation. *In*: Cardiac Electrophysiology and Arrhythmias, ed. D. P. Zipes, J. Jalife. Orlando, Fla.: Grune & Stratton.

Anderson, P.A.V. (1980) Epithelial Conduction: Its Properties and Function. Prog. Neurobiol. 15, 161-203.

Antzelevitch, C., Moe, G. K. (1983) Electrotonic Inhibition and Summation of Impulse Conduction Mammalian Purkinje Fibers. Am. J. Physiol. 245, H42-H53.

Armour, J. A., Sinha, S. (1975) Localized Myocardial Responses to Stimulation of Small Cardiac Branches of the Vagus. J. Physiol. 228, 141-148.

Armour, J. A., Hageman, G. R., Randall, W. C. (1972) Arrhythmias Induced by Local Cardiac Nerve Stimulation. Am. J. Physiol. 223(5), 1068-1075.

Balakhovsky, I. S. (1965) Several Modes of Excitation Movement in Ideal Excitable Tissue. Biophysics 10, 1175-1179.

Baranova, N. B., Mamaev, A. V., Pilipetsky, N. F., Shkunov, V. V., Zel'dovich, B. Ya. (1983) Wave-Front Dislocations: Topological Limitations for Adaptive Systems with Phase Conjugation. J. Opt. Soc. Am. 73, 525-528.

Bardy, G. H., Ungerleider, R. M., Smith, W. M., Ideker, R. E. (1983) A Mechanism of Torsade des Pointes in a Canine Model. Circ. 67(1), 52-59.

Battersby, E. J. (1965) Pacemaker Periodicity in Atrial Fibrillation. Circ. Res. 17, 296-302.

Bauer, W. R., Crick, F.H.C., White, J. H. (1979) Supercoiled DNA. Sci. Am. 243(7), 118-134.

Beeler, G. W., Reuter, H. (1977) Reconstruction of the Action Potential of Ventricular Myocardial Fibres. J. Physiol. 268, 177-210.

Bellet, S. (1971) Clinical Disorders of the Heartbeat. Philadelphia: Lea and Febiger.

Belousov, B. P. (1959) Oscillation Reaction and Its Mechanism (in Russian). Sbornik Referatov po Radiacioni Medicine, p. 145. 1958 Meeting.

Belousov, B. P. (1985) A Periodic Reaction and Its Mechanism. *In*: Oscillations and Traveling Waves in Chemical Systems, ed. R. Field, M. Burger. New York: Wiley.

Bernal, J. D. (1965) Molecular Structure, Biochemical Function, and Evolution. *In*: Theoretical and Mathematical Biology, ed. T. H. Waterman, H. J. Morowitz, p. 96. New York: Blaisdell Publishing Co.

Bernstein, H. J., Phillips, A. V. (1981) Fiber Bundles and Quantum Theory. Sci. Am. 246(7), 123-137.

Berry, M. V. (1980) Some Geometric Aspects of Wave Motion: Wavefront

Dislocations, Diffraction Catastrophies, Diffractals. *In*: The Geometry of the Laplace Operator, ed. R. Osserman, A. Weinstein. Proc. Symp. Pure Maths. 36, 18-28.

Berry, M. V. (1981) Singularities in Waves and Rays. *In*: Les Houches Lectures 35, 453-543. Ed. R. Bahian, M. Kleman, and J. P. Poirer. Amsterdam: North-Holland.

Best, E. N. (1976) Null Space and Phase Resetting Curves for the Hodgkin-Huxley Equations. Ph.D. thesis, Purdue University.

Best, E. N. (1979) Null Space in the Hodgkin-Huxley Equations: A Critical Test. Biophys. J. 27, 87-104.

Bleeker, W. K. (1982) Structure and Function of the Rabbit Sinus Node. Thesis, University of Amsterdam.

Boineau, J., Schuessler, R., Mooney, C., Wylds, A., Miller, C., Hudson, R., Borremans, J., Brockus, C. (1978) Multicentric Origin of the Atrial Depolarization Wave: The Pacemaker Complex. Circ. 58, 1036-1048.

Boineau, J., Schuessler, R., Mooney, C., Miller, C., Wylds, A., Hudson, R., Borremans, J., Brockus, C. (1980a) Natural and Evoked Atrial Flutter Due to Circus Movement in Dogs. Am. J. Cardiol. 45, 1167-1181.

Boineau, J., Schuessler, R., Hackel, D., Miller, C., Brockus, C., Wylds, A. (1980b) Widespread Distribution and Rate Differentiation of the Atrial Pacemaker Complex. Am. J. Physiol. 239, H406-H415.

Bolker, E. D. (1973) The Spinor Spanner. Am. Math. Mo. 80, 977-984.

Bonner, J. T. (1967) The Cellular Slime Molds. Princeton, N.J.: Princeton Univ. Press.

Botre, F., D'Ascenzo, E., Lucarini, C., Memoli, A. (1983) A New Approach to Study Oscillating Chemical Reactions. Bioelectrochem. Bioenerg. 11, 89-95.

Bouligand, Y. (1972) Twisted Fibrous Arrangements in Biological Materials and Cholesteric Mesophases. Tissue and Cell 4, 187-217.

Bray, W. C. (1921) A Periodic Reaction in Homogeneous Solution and Its Relation to Catalysis. J. Am. Chem. Soc. 43, 1262-1267.

Briller, S. A. (1966) Electrocution Hazards. *In*: Mechanisms and Therapy of Cardiac Arrhythmias, ed. L. S. Dreifus, W. Likoff, J. H. Moyer. New York: Grune & Stratton.

Brooks, C. McC., Hoffman, B. F., Suckling, E. E., Orias, O. (1955) Excitability of the Heart. New York: Grune & Stratton.

Brown, G. L., Eccles, J. C. (1934a) The Action of a Single Vagal Volley on the Rhythm of the Heart Beat. J. Physiol. 82, 211-241.

Brown, G. L., Eccles, J. C. (1934b) Further Experiments on Vagal Inhibition of the Heart Beat. J. Physiol. 82, 242-257.

Brown, H. F., Kimura, J., Noble, D., Noble, S. J., Taupignon, A. (1984a)

The Slow Inward Current, i_{si}, in the Rabbit Sino-atrial Node Investigated by Voltage Clamp and Computer Simulation. Proc. Roy. Soc. Lond. B222, 305-328.

Brown, H. F., Kimura, J., Noble, D., Noble, S. J., Taupignon, A. (1984b) The Ionic Currents Underlying Pacemaker Activity in Sino-atrial Node: Experimental Results and Computer Simulations. Proc. Roy. Soc. Lond. B222, 329-347.

Brusca, A., Rosettani, E. (1973) Activation of the Human Fetal Heart. Am. Heart J. 86, 79-87.

Bryant, P. J., Bryant, S. V., French, V. (1977) Biological Pattern Regeneration and Pattern Formation. Sci. Am. 237(1), 66-81.

Bryant, S. V., French, V., Bryant, P. J. (1981) Distal Regeneration and Symmetry. Science 212, 993-1002.

Buck, J. (1973) Bioluminescent Behavior in Renilla. I. Colonial Responses. Biol. Bull. 144, 19-42.

Burch, G. E., Depasquale, N. P. (1965) Sudden, Unexpected, Natural Death. Am. J. Med. Sci. 249, 86-97.

Busse, H. G. (1969) A Spatial Periodic Homogeneous Chemical Reaction. J. Phys. Chem. 73, 750.

Cardinal, R., Savard, P., Carson, L., Perry, J.-B., Pagge, P. (1984) Mapping of Ventricular Tachycardia Induced by Programmed Stimulation in Canine Preparations of Myocardial Infarction. Circ. 70(1), 136-148.

Castellanos, A., Luceri, R. M., Moliero, F., Kayden, D. S., Trohman, R. G., Zaman, L. (1984) Annihilation, Entrainment, and Modulation of Ventricular Parasystolic Rhythms. Am. J. Cardiol. 54, 317-322.

Chay, T. R., Lee, Y. S. (1984) Impulse Responses of Automaticity in the Purkinje Fiber. Biophys. J. 45, 871-849.

Chay, T. R., Lee, Y. S. (1985) Phase Resetting and Bifurcation in the Ventricular Myocardium. Biophys. J. 47(5), 641-651.

Chen, P.-S., Shibata, N., Dixon, E. G., Martin, R. O., Ideker, R. E. (1986a) Comparison of the Defibrillation Threshold and the Upper Limit of Ventricular Vulnerability. Circ. 73(5):1022-1028.

Chen, P.-S., Shibata, N., Dixon, E. G., Wolf, P., Daniely, N., Sweeney, M., Smith, W., Ideker, R. E. (1986b) Activation during Ventricular Defibrillation in Open Chest Dogs: Evidence of Complete Cessation and Regeneration of Ventricular Fibrillation. J. Clin. Invest. 77, 810-823.

Clay, J. R., Guevara, M., Shrier, A. (1984) Phase Resetting of the Rhythmic Activity of Embryonic Heart Cell Aggregates: Experiment and Theory. Biophys. J. 45, 699-714.

Cohen, D. S., Neu, J. C., Rosales, R. R. (1978) Rotating Spiral Wave

Solutions of Reaction-Diffusion Equations. SIAM J. Appl. Math. 35, 536-547.

Cohn, R. L., Rush, S., Lepeshkin, E. (1982) Theoretical Analyses and Computer Simulation of ECG Ventricular Gradient and Recovery Waveforms. IEEE Trans. Biomed. Eng. 29(6), 413-422.

Cole, K. S. (1968) Membranes, Ions, and Impulses. Berkeley: Univ. of Calif. Press.

Colizza, D., Guevara, M., Shrier, A. (1983) A Comparative Study of Collagenase- and Trypsin-dissociated Embryonic Heart Cells: Reaggregation, Electrophysiology, and Pharmacology. Can. J. Physiol. Pharm. 61, 408-419.

Corr, P. B., Gillis, R. A. (1978) Autonomic Neural Influences on the Dysrhythmias Resulting from Myocardial Infarction. Circ. Res. 43(1), 1-9.

Cranefield, P. F., Hoffman, B. F., Siebens, A. A. (1957) Anodal Excitation in Heart Muscle. Am. J. Physiol. 190(2), 383-390.

Crick, F.H.C. (1976) Linking Numbers and Nucleosomes. Proc. Natl. Acad. Sci. USA 73, 2639-2643.

Daggett, W. M., Wallace, A. G. (1966) Vagal and Sympathetic Influences on Ectopic Impulse Formation. In: Mechanisms and Therapy of Cardiac Arrhythmias, ed. L. S. Dreifus, W. Likhoff, J. H. Moyer. New York: Grune & Stratton.

de Bakker, J.M.T., Henning, B., Merx, W. (1979) Circus Movement in Canine Right Ventricle. Circ. Res. 45, 374-378.

de Bakker, J.M.T., Janse, M. J., van Capelle, F.J.L., Durrer, D. (1983) Endocardial Mapping by Simultaneous Recording of Endocardial Electrograms during Cardiac Surgery for Ventricular Aneurysm. J. Am. Coll. Cardiol. 2(5), 947-953.

de Bruin, G. D., Ypey, D. L., Vanmeerwijk, W.P.M. (1984) Synchronization in Chains of Pacemaker Cells by Phase Resetting Action Potential Effects. Biol. Cybern. 48, 176-186.

de Vega, H. J. (1978) Closed Vortices and the Hopf Index in Classical Field Theory. Phys. Rev. D 18, 2945-2951.

Delbruck, M. (1962) Knotting Problems in Biology. Mathematical Problems in the Biological Sciences. Proc. Symp. Appl. Math. 14, 55-63. Providence R.I.: A.M.S.

Denes, P., Gabster, A., Huang, S. K. (1981) Clinical, Electrocardiographic, and Follow-Up Observations in Patients Having Ventricular Fibrillation During Holter Monitoring. Am. J. Cardiol. 48, 9-16.

DeSilva, R. (1982) Central Nervous System Risk Factors for Sudden Cardiac Death. Ann. N.Y. Acad. Sci. 382, 143-161.

DeSimone, J. A., Beil, D. L., Scriven, L. E. (1973) Ferroin-Collodion

Membranes: Dynamic Concentration Patterns in Planar Membranes. Science 180, 946-948.

Dharmananda, S. (1980) Studies of the Circadian Clock of Neurospora crassa: Light-Induced Phase Shifting. Ph.D. thesis, Univ. of Calif., Santa Cruz.

Dharmananda, S., Feldman, J. F. (1979) Spatial Distribution of Circadian Rhythm in Aging Cultures of Neurospora crassa. Plant Physiol. 63, 1049-1054.

Diaz, P. J., Rudy, Y., Plonsey, R. (1984) Intercalated Discs as a Cause for Discontinuous Propagation in Cardiac Muscle: A Theoretical Simulation. Ann. Biomed. Eng. 11, 177-189.

Dillon, S., Morad, M. (1981) A New Laser Scanning System for Measuring Action Potential Propagation in the Heart. Science 214, 453-456.

Dillon, S., Ursell, P. C., Wit, A. L. (1985) Pseudo-block Caused by Anisotropic Conduction: A New Mechanism for Sustained Reentry. Circ. 72 Suppl. III, 1116.

Dolnik, M., Schreiber, I., Marek, M. (1984) Experimental Observations of Periodic and Chaotic Regimes in a Forced Chemical Oscillator. Physics Letters 100a(6), 316-319.

Dolnik, M., Schreiber, I., Marek, M. (1986) Dynamic Regimes in Periodically Forced Reaction Cell with Oscillatory Chemical Reaction. Physica D 21(1), 78-92.

Donders, F. C. (1868) Zur Physiologie des Nervus Vagus. Pflugers Archiv 1, 331-361.

Downar, E., Parson, I. D., Mickleborough, L. L., Yao, L. C., Cameron, D. A., Waxman, M. B. (1984) On-Line Epicardial Mapping of Intraoperative Ventricular Arrhythmias: Initial Clinical Experience. J. Am. Coll. Cardiol. 4(4), 703-714.

Dulos, E. (1981) Synchronization of Chemical Oscillation by Periodic Light Pulses. In: Non-Linear Phenomena in Chemical Dynamics, ed. C. Vidal, A. Pacault, pp. 140-146. Berlin: Springer-Verlag.

Dulos, E., De Kepper, P. (1983) Experimental Study of Synchronization Phenomena under Periodic Light Irradiation of a Nonlinear Chemical System. Phys. Chem. 18, 211-223.

Durrer, D., van Dam, R. T., Freud, G. E., Janse, M. J., Meijler, F. L., Arzbaecher, R. C. (1970) Total Excitation of the Isolated Human Heart. Circ. 41, 899-912.

Durston, A. J. (1973) Dictyostelium Discoideum Aggregation Fields as Excitable Media. J. Theor. Biol. 42, 483-504.

Ebihara, L., Johnson, E. A. (1980) Fast Sodium Current in Cardiac Muscle: A Quantitative Description. Biophys. J. 32, 779-790.

Eccles, J. C., Hoff, H. E. (1934) The Rhythm of the Heart Beat. I, II, and III. Proc. Roy. Soc. Lond. 115, 307-369.

El-Sherif, N. (1985) The Figure 8 Model of Reentrant Excitation in the Canine Postinfarction Heart. *In*: Cardiac Electrophysiology and Arrhythmias, ed. D. P. Zipes, J. Jalife, pp. 363-378. Orlando, Fla.: Grune & Stratton.

El-Sherif, N., Hope, R. R., Scherlag, B. J., Lazzara, R. (1977) Re-Entrant Ventricular Arrhythmias in the Late Myocardial Infarction Period. 2. Patterns of Initiation and Termination of Re-Entry. Circ. 55, 702-719.

El-Sherif, N., Mehra, R., Gough, W. B., Zeiler, R. H. (1982) Ventricular Activation Patterns of Spontaneous and Induced Ventricular Rhythms in Canine One-Day-Old Myocardial Infarction. Circ. Res. 51, 152-166.

El-Sherif, N., Gough, W. B., Zeiler, R. H., Hariman, R. (1985) Reentrant Ventricular Arrhythmias in the Late Myocardial Infarction Period. 12. Spontaneous vs. Induced Reentry and Intramural vs. Epicardial Circuits. J. Am. Coll. Cardiol. 6(1), 124-132.

Elsdale, T., Wasoff, F. (1976) Fibroblast Cultures and Dermatoglyphics: The Topology of Two Planar Patterns. Wilhelm Roux' Archiv. Entwick. Org. 180, 121-147.

Engel, G. L. (1978) Psychologic Stress, Vasodepressor (vasovagal) Syncope, and Sudden Death. Ann. Int. Med. 89, 403-412.

Engelmann, W., Johnsson, A. (1978) Attenuation of the Petal Movement Rhythm in Kalanchoë with Light Pulses. Physiol Plant. 43, 68-76.

Enright, J. T. (1980) The Timing of Sleep and Wakefulness. Berlin: Springer-Verlag.

Enright, J. T., Winfree, A. T. (1987) Detecting a Phase Singularity in a Coupled Stochastic System. *In*: Modeling Circadian Rhythms, ed. G. Carpenter. Some Mathematical Questions in Biology. Providence, R.I.: A.M.S.

Epstein, I. R., Kustin, K., De Kepper, P., Orban, M. (1983) Oscillating Chemical Reactions. Sci. Am. 248(3), 112-123.

Farber, I. C., Grinvald, A. (1982) A Novel Tool for Evaluation of Neural Networks. Neurosci. Abstr. 8(2), 683.

Farley, B. G. (1965) A Neural Network Model and the Slow Potentials of Electrophysiology. Computers in Biomed. Res. 1, 265-294.

Fel'd, B. N., Morozova, O. L. (1979) Role of the Mutual Electrical Influence of Adjacent Portions of the Myocardium in the Genesis of Arrhythmias. Biophysics 23, 891-897.

Ferrier, G. R., Rosenthal, J. E. (1980) Automaticity and Entrance Block Induced by Focal Depolarization of Mammalian Ventricular Tissues. Circ. Res. 47, 238-248.

Ferrier, G. R., Saunders, J. H., Mendez, C. (1973) A Cellular Mechanism for the Generation of Ventricular Arrhythmias by Acetylstrophanthidin. Circ. Res. 32, 600-609.

Ferris, L. P., King, B. G., Spence, P. W., Williams, H. B. (1936) Effect of Electric Shock on the Heart. Electrical Engineering 55, 489-515.

Field, R., Burger, M., eds. (1985) Oscillations and Traveling Waves in Chemical Systems. New York: Wiley.

Fife, P. C. (1984) Propagator-Controller Systems and Chemical Patterns. In: Non-Equilibrium Dynamics in Chemical Systems, ed. C. Vidal, A. Pacault. Berlin: Springer-Verlag.

FitzHugh, R. (1960) Thresholds and Plateaus in the Hodgkin-Huxley Nerve Equations. J. Gen. Physiol. 43, 867-896.

FitzHugh, R. (1961) Impulses and Physiological States in Theoretical Models of Nerve Membrane. Biophys. J. 1, 445-466.

Forster, F. K., Weaver, W. D. (1982) Recognition of Ventricular Fibrillation, Other Rhythms, and Noise in Patients Developing the Sudden Cardiac Death Syndrome. Computers in Cardiol. 1982, 245-248.

Foy, J. L. (1974) A Computer Simulation of Impulse Conduction in Cardiac Muscle. Technical Report no. 166, Univ. of Mich. Dept. of Computer Science.

Fuller, F. B. (1971) The Writhing Number of a Space Curve. Proc. Natl. Acad. Sci. USA 68, 815-819.

Fuller, F. B. (1978) Decomposition of the Linking Number of a Closed Ribbon: A Problem from Molecular Biology. Proc. Natl. Acad. Sci. USA 75, 3557-3561.

Gardner, M. (1970) Conway's Game of Life. Sci. Am. 223(10), 120-123 (Mathematical Games), and followup in 1971, 224(2), 104-108.

Garrey, W. E. (1914) The Nature of Fibrillary Contraction of the Heart— Its Relation to Tissue Mass and Form. Am. J. Physiol. 33, 397-414.

Garrey, W. E. (1924) Auricular Fibrillation. Physiol. Rev. 4, 215-250.

Gerisch, G. (1965) Stadienspezifische Aggregationsmuster bei Distyostelium Discoideum. Wilhelm Roux' Archiv. Entwick. Org. 156, 127-144.

Geuze, R. H., Koster, R. W. (1984) Ventricular Fibrillation and Transient Arrhythmias after Defibrillation in Patients with Acute Myocardial Infarction. J. Electrocard. 17(4), 353-360.

Ghosh, A., Chance, B. (1964) Metabolic Coupling and Synchronization of NADH Oscillations in Yeast Cell Populations. Biophys. Res. Comm. 16, 174-181.

Gilmour, R. F., Heger, J. J., Prystowsky, E. N., Zipes, D. P. (1983) Cellular Electrophysiologic Abnormalities of Diseased Human Ventricular Myocardium. Am. J. Physiol. 51(11), 137-144.

Glass, L. (1977) Patterns of Supernumerary Limb Regeneration. Science 198, 321-322.

Glass, L., Winfree, A. T. (1984) Discontinuities in Phase-Resetting Experiments. Am. J. Physiol. 246, R251-258.

Gomatam, J. (1982) Pattern Synthesis from Singular Solutions in the Debye Limit: Helical Waves and Twisted Toroidal Scroll Structures. J. Phys. a15, 1463-1476.

Gordon, H., Winfree, A. T. (1978) A Single-Spiral Artifact in Arthropod Cuticle. Tissue and Cell 10, 39-50.

Gorelova, N. A., Bures, J. (1983) Spiral Waves of Spreading Depression in the Isolated Chicken Retina. J. Neurobiol. 14(5), 353-363.

Gradman, A. H., Bell, P. A., Debusk, R. F. (1978) Sudden Death during Ambulatory Monitoring: Clinical and Electrocardiographic Correlations Report of a Case. Circ. 55, 210-211.

Greenberg, J. M., Hassard, B. D., Hastings, S. P. (1978) Pattern Formation and Periodic Structure in Systems Modeled by Reaction-Diffusion Equations. Bull. Am. Math. Soc. 84, 1296-1327.

Grenader, A. K., Zurabishvily, G. G. (1984) Different Sensitivities of Two Mechanisms of Excitation Wave Circulation in Heart Tissue to Antiarrhythmic Blockers of Fast Sodium Current (in Russian). Biofizika 30(1), 118-123.

Guckenheimer, J. (1975) Isochrons and Phaseless Sets. J. Math. Biol. 1, 259-273.

Guevara, M. (1984) Chaotic Cardiac Dynamics. Ph.D. thesis, McGill University.

Guevara, M., Glass, L. (1982) Phase Locking, Period Doubling Bifurcations, and Chaos in a Mathematical Model of a Periodically Driven Biological Oscillator. J. Math. Biol. 14, 1-23.

Guevara, M., A. Shrier, and L. Glass (1986) Phase Resetting of Spontaneously Beating Embryonic Ventricular Heart-Cell Aggregates. Am. J. of Physiol. In press.

Gul'ko, F. B., Petrov, A. A. (1972) Mechanism of the Formation of Closed Pathways of Conduction in Excitable Media (in Russian). Biofizika 17, 261-270.

Guttman, R., Lewis, S., Rinzel, J. (1980) Control of Repetitive Firing in Squid Axon Membrane as a Model for a Neurone Oscillator. J. Physiol. 305, 377-395.

Hagan, P. S. (1982) Spiral Waves in Reaction-Diffusion Equations. SIAM J. Appl. Math. 42, 762-786.

Hageman, G. R., Goldberg, J. M., Armour, J. A., Randall, W. C. (1973) Cardiac Dysrhythmias Induced by Autonomic Nerve Stimulation. Am. J. Cardiol. 32(11), 823-830.

Han, J., Moe, G. K. (1964) Nonuniform Recovery of Excitability in Ventricular Muscle. Circ. Res. 14(1), 44-60.

Han, J., Garcia De Jalon, P., Moe, G. K. (1964) Adrenergic Effects on Ventricular Vulnerability. Circ. Res. 14(6), 516-524.

Harumi, K., Smith, C. R., Abildskov, J. A., Burgess, M. J., Lux, R. L., Wyatt, R. F. (1980) Detailed Activation Sequence in the Region of Electrically Induced Ventricular Fibrillation in Dogs. Jap. Heart. J. 21, 533-544.

Hastings, S. (1981) Persistent Spatial Patterns for Semi-Discrete Models of Excitable Media. J. Math. Biol. 11, 105-117.

Helmholtz, H. von (1867) On Integrals of the Hydrodynamic Equations Which Express Vortex Motions. Phil. Mag. Suppl. 33, 485-510 (translation of 1858 German article).

Herbschleb, J. N., Heethaar, R. M., van der Tweel, I., Zimmerman, A.N.E., Meijler, F. L. (1979) Signal Analysis of Ventricular Fibrillation. Computers in Cardiol. 1979, 49-52.

Herbschleb, J. N., Heethaar, R. M., van der Tweel, I., Meijler, F. L. (1980) Frequency Analysis of the ECG Before and During Ventricular Fibrillation. Computers in Cardiol. 1980, 365-368.

Herbschleb, J., van der Tweel, I., Meijler, F. (1982) The Apparent Repetition Frequency of Ventricular Fibrillation. Computers in Cardiol. 1982, 249-252.

Herbschleb, J. N., van der Tweel, I., Meijler, F. (1983) The Illusion of Travelling Wavefronts during Ventricular Fibrillation. Circ. 68, Supp. III, 343.

Hille, B. (1984) Ionic Channels of Excitable Membranes. Sunderland, Mass.: Sinauer.

Hinkle, L. E., Argyros, K. C., Hayes, J. C., Robinson, T., Alonso, D. R. (1977) Pathogenesis of an Unexpected Sudden Death: Role of Early Cycle Ventricular Premature Contractions. Cardiol. 39, 873-879.

Hirano, Y., Sawanobori, T., Hiraoka, M. (1982) Circus Movement Tachycardia Examined by the Optical Recording and by the Computer Simulation. Jap. Heart J. 23, 109-114.

Hohnloser, S., Weiss, M., Zeiher, A., Wollschlager, H., Hust, M. H., Just, H. (1984) Sudden Cardiac Death Recorded during Ambulatory Electrocardiographic Monitoring. Clin. Cardiol. 7, 517-523.

Honerkamp, J. (1983) The Heart as a System of Coupled Nonlinear Oscillators. J. Math. Biol. 18(1), 69-88.

Hoppensteadt, F. C. (1985) Introduction to the Mathematics of Neurons. Cambridge: Cambridge Univ. Press.

Hoppensteadt, F. C., Keener, J. P. (1982) Phase Locking of Biological Clocks. J. Math. Biol. 15(3), 339-349.

Horton, J. C. (1984) Cytochrome Oxidase Patches: A New Cytoarchi- tectonic Feature of Monkey Visual Cortex. Phil. Trans. Roy. Soc. B. 304, 199-253.

Horton, J. C., Hedley-White, E. T. (1984) Mapping of Cytochrome Oxidase Patches and Ocular Dominance Columns in Human Visual Cortex. Phil. Trans. Roy. Soc. B. 304, 255-272.

Ideker, R. E., Klein, G. J., Harrison, L., Smith, W. M., Kasell, J., Reimer, K. A. (1981) The Transition to Ventricular Fibrillation Induced by Reperfusion after Acute Ischemia in the Dog: A Period of Organized Epicardial Activation. Circ. 63, 1371-1379.

Ideker, R. E., Bardy, G. H., Worley, S. J., German, L. D., Smith, W. M. (1984) Patterns of Activation during Ventricular Fibrillation. In: Tachycardias: Mechanisms, Diagnoses, and Treatment, ed. M. E. Josephson, H.J.J. Wellens, pp. 519-536. Philadelphia: Lea and Febirger.

Ideker, R. E., Shibata, N. (1986) Experimental Evaluation of Predictions about Ventricular Fibrillation based upon Phase Resetting. Proc. Int. Union of Physiol. Sci., 16, 88 (Abstract 165.01).

Ivanitsky, G. R., Krinsky, V. I., Zaikin, A. N., Zhabotinsky, A. M. (1981) Autowave Processes and Their Role in Disturbing the Stability of Distributed Excitable Systems. Soviet Scientific Reviews D 2, 280-324.

Jalife, J. (1984) Mutual Entrainment and Electrical Coupling as Mechanisms for Synchronous Firing of Rabbit Sino-Atrial Pacemaker Cells. J. Physiol. Lond. 356, 221-243.

Jalife, J., Antzelevitch, C. (1979) Phase Resetting and Annihilation of Pacemaker Activity in Cardiac Tissue. Science 206, 696-697.

Jalife, J., Antzelevitch, C. (1980) Pacemaker Annihilation: Diagnostic and Therapeutic Implications. Am. Heart J. 100, 128-130.

Jalife, J., Michaels, D. C. (1985) Phase-Dependent Interactions of Cardiac Pacemakers as Mechanisms of Control and Synchronization in the Heart. In: Cardiac Electrophysiology and Arrhythmias, ed. D. Zipes, J. Jalife, pp. 109-122. Orlando, Fla.: Grune & Stratton.

Jalife, J., Slenter, V.A.J., Salata, J. J., Michaels, D. C. (1983) Dynamic Vagal Control of Pacemaker Activity in the Mammalian Sinoatrial Node. Circ. Res. 52, 642-656.

Janse, M., van Capelle, F. (1982) Electrotonic Interactions across an Inexcitable Region as a Cause of Ectopic Activity in Acute Regional Myocardial Ischemia. Circ. Res. 50, 527-537.

Janse, M. J., van Capelle, F. J., Morsink, H., Kleber, A. G., Wilms-Schopman, F., Cardinal, R., et al. (1980) Flow of Injury Currents and Patterns of Excitation during Early Ventricular Arrhythmias in

Acute Regional Myocardial Ischemia in Isolated Hearts. Circ. Res. 47, 151-165.

Johnsson, A. (1976) Oscillatory Transpiration and Water Uptake of Avena Plants. Bull. Inst. Math. Appl. 12, 22-26.

Johnsson, A., Karlsson, H. G. (1971) A Feedback Model for Biological Rhythms. I. Mathematical Description and Basic Properties of the Model. *In*: Proceedings of the First Euro. Biophys. Congr., ed. E. Broda, A. Locker, pp. 263-267. Vienna: Academy of Medicine Press.

Jongsma, H. J., van Rijn, H. E. (1972) Electrotonic Spread of Current in Monolayer Cultures of Neonatal Rat Heart Cells. J. Membr. Biol. 9, 341-360.

Jongsma, H. J., Tsjernina, L., Debruijne, J. (1983) The Establishment of Regular Beating in Populations of Pacemaker Heart Cells: A Study with Tissue-Cultured Rat Heart Cells. J. Mol. Cell. Cardiol. 15, 123-133.

Josephson, R. K., Schwab, W. E. (1979) Electrical Properties of an Excitable Epithelium. J. Gen. Physiol. 74, 213-236.

Josephson, M. E., Spielman, S. R., Greenspan, A. M., Horowitz, L. N. (1979) Mechanism of Ventricular Fibrillation in Man. Am. J. Cardiol. 44, 623-631.

Joyner, R. W., Picone, J., Veenstra, R., Rawling, D. (1983) Propagation through Electrical Coupled Cells: Effects of Regional Changes in Membrane Properties. Circ. Res. 53, 526-534.

Kannel, W., Thomas, H. (1982) Sudden Coronary Death: The Framingham Study. Ann. N.Y. Acad. Sci. 382, 3-21.

Katz, L. N. (1946) Electrocardiography. Philadelphia: Lea and Febiger.

Katz, L. N., Pick, A. (1956) Clinical Electrocardiography I: Arrhythmias. Philadelphia: Lea and Febiger.

Kauffman, S. A., Wille, J. J. (1975) The Mitotic Oscillator in Physarum Polycephalum. J. Theor. Biol. 55, 47-93.

Kawato, M. (1981) Transient and Steady State Phase Response Curves of Limit Cycle Oscillators. J. Math. Biol. 12, 13-30.

Kawato, M., Suzuki, R. (1978) Biological Oscillators Can Be Stopped: Topological Study of a Phase Response Curve. Biol. Cybernetics 30, 241-248.

Keener, J. P. (1986) A Geometrical Theory for Spiral Waves in Excitable Media. SIAM J. Appl. Math. In press.

Keener, J. P., Glass, L. (1984) Global Bifurcations of a Periodically Forced Nonlinear Oscillator. J. Math. Biol. 21(2), 175-190.

Keener, J. P., Tyson, J. J. (1986) Spiral Waves in the Belousov-Zhabotinsky Reaction. Physica D. In press.

Kempf, F. C., Josephson, M. E. (1984) Cardiac Arrest Recorded on Ambulatory Electrocardiograms. Am. J. Cardiol. 53, 1577-1582.

Kibble, T.W.B. (1976a) Some Implications of a Cosmological Phase Transition. Physics Reports 67(1), 183-199.

Kibble, T.W.B. (1976b) Topology of Cosmic Domains and Strings. J. Phys. A.: Math. Gen. 9(8), 1387-1398.

Kida, S. (1981) A Vortex Filament Moving without Change of Form. J. Fluid Mech. 112, 397-409.

Kida, S. (1982) Stability of a Steady Vortex Filament. J. Phys. Soc. Japan 51, 1655-1662.

King, B. G. (1934) The Effect of Electric Shock on Heart Action with Special Reference to Varying Susceptibility in Different Parts of the Cardiac Cycle. Thesis, Columbia University. Aberdeen Press.

Koga, S. (1982) Rotating Spiral Waves in Reaction-Diffusion Systems: Phase Singularities of Multi-Armed Waves. Prog. T. Phys. 67, 164-178.

Kogan, B. Y., Zykov, V. S., Petrov, A. A. (1980) Hybrid Computer Simulation of Stimulative Media. *In*: Simulation of Systems '79, ed. L. Dekker, G. Savastano, G. C. Vansteenkiste, pp. 693-701. Amsterdam: North-Holland.

Korner, P. I. (1979) Central Nervous Control of Autonomic Cardiovascular Function. *In*: Handbook of Physiology, section 2, vol. 1, ed. R. M. Berne et al. Bethesda: Am. Physiol. Soc.

Kralios, F. A., Martin, L., Burgess, M. J., Millar, K. (1975) Local Ventricular Repolarization Changes due to Nerve-Branch Stimulation. Am. J. Physiol. 228(5), 1621-1626.

Kramer, J. B., Saffitz, J. E., Witkowski, F. X., Corr, P. B. (1985) Intramural Reentry as a Mechanism of Ventricular Tachycardia during Evolving Canine Myocardial Infarction. Circ. Res. 56, 736-754.

Krinsky, V. I. (1966) Spread of Excitation in an Inhomogeneous Medium (State Similar to Cardiac Fibrillation) (in Russian). Biofizika 11, 776-784.

Krinsky, V. I. (1968) Fibrillation in Excitable Media (in Russian). Prob. in Cybern. 20, 59-80.

Krinsky, V. I. (1978) Mathematical Models of Cardiac Arrhythmias (Spiral Waves). Pharmac. Ther. B, 3, 539-555.

Krinsky, V. I., ed. (1984) Self Organization: Autowaves and Structures far from Equilibrium. Berlin: Springer-Verlag.

Krinsky, V. I., Kholopov, A. V. (1967) Echo in Excitable Tissue (in Russian). Biofizika 12, 600-606.

Krinsky, V. I., Malomed, B. A. (1983) Quasi-Harmonic Rotating Waves in Distributed Active Systems. Physica D 9(1-2), 81-95.

Kuo, S., Dillman, R. (1978) Computer Detection of Ventricular Fibrillation. Computers in Cardiol. 1978, 347-349.

Kuramoto, Y. (1984) Chemical Oscillations, Waves, and Turbulence. Berlin: Springer-Verlag.

Lahiri, A., Balasubramian, V., Raftery, E. B. (1979) Sudden Death during Ambulatory Monitoring. Brit. Med. J. 1, 1676-1678.

Lashley, K. S. (1941) Patterns of Cerebral Integration Indicated by the Scotomas of Migraine. Arch. Neurol. Psychiat. 46, 331-339.

Lauritzen, M., Olsen, T. S., Lassen, N. A., Paulson, O. B. (1983) Changes in Regional Cerebral Blood Flow during the Course of Classic Migraine Attacks. Ann. Neurol. 13, 633-641.

Lewis, T. (1925) The Mechanism and Graphic Registration of the Heart Beat, 3rd ed. London: Shaw and Sons Ltd.

Lown, B. (1979a) Sudden Cardiac Death. Circ. 60, 1593-1599.

Lown, B. (1979b) Sudden Cardiac Death: The Major Challenge Confronting Contemporary Cardiology. Am. J. Cardiol. 43, 313-328.

Lown, B., Verrier, R. L. (1976) Neural Activity and Ventricular Fibrillation. N. Engl. J. Med. 294, 1165-1170.

Lown, B., Verrier, R. L., Rabinowitz, S. H. (1977) Neural and Psychologic Mechanisms and the Problem of Sudden Cardiac Death. Am. J. Cardiol. 39, 890-902.

McAllister, R. E., Noble, D., Tsien, R. W. (1975) Reconstruction of the Electrical Activity of Cardiac Purkinje Fibers. J. Physiol. 251, 1-59.

MacGregor, R. J., Lewis, E. R. (1977) Neural Modelling. New York: Plenum.

McMahon, T. A., Bonner, J. T. (1983) On Size and Life. Sci. Am. Library. New York: W. H. Freeman and Co.

MacWilliam, J. A. (1888) On the Effects of Increased Arterial Pressure on the Mammalian Heart. Proc. Roy. Soc. Lond. 44, 287-292.

Madore, B. F., Freedman, W. L. (1983) Computer Simulations of the Belousov-Zhabotinsky Reaction. Science 222, 615-616.

Maki, K. (1978) Textures in Superfluid 3He. In: Solitons and Condensed Matter Physics, ed. A. R. Bishop, T. Schneider, pp. 278-290. Berlin: Springer-Verlag.

Malchow, D., Nanjundiah, V., Gerisch, G. (1978) Ph Oscillations in Cell Suspensions of Dictyostelium Discoideum: Their Relation to Cyclic Amp Signals. J. Cell. Sci. 30, 319-330.

Malinowski, J. R., Laval-Martin, D. L., Edmunds, L. N., Jr. (1985) Circadian Oscillators, Cell Cycles, and Singularities: Light Perturbations of the Free-runing Rhythm of Cell Division in Euglena. J. Comp. Physiol. B 155, 257-276.

Malomed, B. A., Rudenko, A. N. (1986) The Approximate Analytical

Theory of Spiral and Concentric Waves in Active Media. Physica D. In press.

Mandel, W. J., ed. (1980) Cardiac Arrhythmias. Philadelphia: J. B. Lippincott Co.

Martins, J. B., Zipes, D. P. (1980) Effects of Sympathetic and Vagal Nerves on Recovery Properties of the Endocardium and Epicardium of the Canine Left Ventricle. Circ. Res. 46, 100-110.

Matta, R. J., Verrier, R. L., Lown, B. (1979) Repetitive Extrasystole as an Index of Vulnerability to Ventricular Fibrillation. Am. J. Physiol. 230, 1469-1473.

Mayer, A. G. (1906) Rhythmical Pulsation in Scyphomedusae. Carnegie Inst. of Wash. Publ. #47.

Medvinsky, A. B., Pertsov, A. M. (1979) Interaction of Fibres on Spread of Excitation in Smooth Muscle and Myocardial Tissues: Electrotonic Interaction. Biophysics 24, 139-145.

Medvinsky, A. B., Pertsov, A. M. (1982) Initiation Mechanism of the First Extrasystole in a Short-Lived Atrial Arrhythmia (in Russian). Biofizika 27, 895-899.

Medvinsky, A. B., Pertsov, A. M., Polishuk, G. A., Fast, V. G. (1983) Mapping of Vortices in Myocardium. In: Electrical Field of the Heart (in Russian), ed. O. Baum, M. Roschevsky, L. Titomir, pp. 38-51. Moscow: Nauka.

Medvinsky, A. B., Panfilov, A. V., Pertsov, A. M. (1984) Properties of Rotating Waves in Three Dimensions: Scroll Rings in Myocardium. In: Self Organization: Autowaves and Structures far from Equilibrium, ed. V. I. Krinsky. Berlin: Springer-Verlag.

Mehra, R., Zeiler, R. H., Gough, W. B., El-Sherif, N. (1983) Reentrant Ventricular Arrhythmias in the Late Myocardial Infarction Period. 9. Electrophysiologic-Anatomic Correlation of Reentrant Circuits. Circ. 67(1), 11-24.

Meinhardt, H. (1982) Models of Biological Pattern Formation. London: Academic Press.

Mermin, N. D. (1979) The Topological Theory of Defects in Ordered Media. Rev. Mod. Phys. 51(3), 591-648.

Michaels, D. C., Matyas, E. P., Jalife, J. (1984) A Mathematical Model of the Effects of Acetylcholine Pulses on Sinoatrial Pacemaker Activity. Circ. Res. 55, 89-101.

Michel, L. (1980) Symmetry Defects and Broken Symmetries: Configurations, Hidden Symmetries. Rev. Mod. Phys. 52(3), 617-651.

Mikhailov, A. S., Krinsky, V. I. (1983) Rotating Spiral Waves in Excitable Media: The Analytic Results. Physica 9D, 346-371.

Mines, G. R. (1914) On Circulating Excitations in Heart Muscles and

Their Possible Relation to Tachycardia and Fibrillation. Trans. Roy. Soc. Can. 4, 43-53.

Miura, R. M., Plant, R. D. (1981) Rotating Waves in Models of Excitable Media. *In*: Differential Equations and Applications in Ecology, Epidemiology, and Population Problems, ed. S. N. Busenberg. New York: Academic Press.

Moe, G. K. (1962) On the Multiple Wavelet Hypothesis of Atrial Fibrillation. Arch. Int. Pharmacodyn. 140, 183-188.

Moe, G. K., Abildskov, J. A. (1959) Atrial Fibrillation as a Self-Sustaining Arrhythmia Independent of Focal Discharge. Am. Heart J. 58, 59-70.

Moe, G. K., Harris, A. S., Wiggers, C. J. (1941) Analysis of the Initiation of Fibrillation by Electrographic Studies. Am. J. Physiol. 134, 473-492.

Moe, G. K., Rheinboldt, W. C., Abildskov, J. A. (1964) A Computer Model of Atrial Fibrillation. Am. Heart J. 67, 200-220.

Moffatt, H. K. (1969) The Degree of Knottedness of Tangled Vortex Lines. J. Fluid Mech. 35, 117-129.

Morgan, J. S. (1916) The Periodic Evolution of Carbon Dioxide. J. Chem. Soc. Trans. 109, 274-279.

Müller, S. C., Plesser, T., Hess, B. (1985) The Structure of the Core of the Spiral Wave in the Belousov-Zhabotinsky Reagent. Science 230, 661-663.

Nakashima, H., Perlman, J., Feldman, J. F. (1981) Cycloheximide-Induced Phase Shifting of the Circadian Clock of Neurospora. Am. J. Physiol. 241, R31-R35.

Nandapurkar, P., Winfree, A. T. (1986) Unpublished computations.

Nash, C., Sen, S. (1983) Topology and Geometry for Physicists. San Diego: Academic Press.

Newell, P. C. (1983) Attraction and Adhesion in the Slime Mold Dictyostelium. *In*: Fungal Differentiation, ed. J. Smith, pp. 43-59. New York: M. Dekker.

Nicolis, G., Prigogine, I. (1977) Self-Organization in NonEquilibrium Systems: From Dissipative Structures to Order through Fluctuations. New York: Academic Press.

Ninomiya, I. (1966) Direct Evidence of Nonuniform Distribution of Vagal Effects on Dog Atria. Circ. Res. 19, 576-583.

Noble, D. (1984) The Surprising Heart: A Review of Recent Progress in Cardiac Electrophysiology. J. Physiol. 353, 1-10.

Noble, D., Noble, S. J. (1984) A Model of Sino-atrial Node Electrical Activity Based on a Modification of the DiFrancesco-Noble (1984) Equations. Proceedings of the Royal Society of London B222, 295-304.

Nye, J. F., Berry, M. V. (1974) Dislocations in Wave Trains. Proc. Roy. Soc. Lond. A336, 165-190.

Nygards, M. E., Hulting, J. (1977) Recognition of Ventricular Fibrillation Utilizing the Power Spectrum of the ECG. Computers in Cardiol. 1977, 393-397.

Odell, G. M., Bonner, J. T. (1986) How the Dictyostelium Discoideum Grex Crawls. In press.

Oster, G. F. (1984) On the Crawling of Cells. J. Embry. Exp. Morph. 83, 327-364.

Oster, G. F., Odell, G. M. (1984a) Mechanics of Cytogels 1: Oscillations in Physarum. Cell Motil. 4(6), 469-503.

Oster, G. F., Odell, G. M. (1984b) Mechanochemistry of Cytogels. Physica 12D, 333-350.

Oster, G. F., Murray, J. D., and Harris, A. K. (1983) Mechanical Aspects of Mesenchymal Morphogenesis. J. Embry. Exp. Morph. 78, 83-125.

Oster, G. F., Murray, J. D., Nani, P. K. (1985) A Model for Chondrocyte Condensations in the Developing Limb: The Role of the Extracellular Matrix and Cell Tractions. J. Embry. Exp. Morph. 89, 93-112.

Ostwald, W. (1900) Periodische Erscheinungen bei der Auflösung des Chrom in Säuren. Zeit. Phys. Chem. 35, 33-76 and 204-256.

Palmer, D. G. (1962) Interruption of T Waves by Premature QRS Complexes and the Relationship of This Phenomenon to Ventricular Fibrillation. Am. Heart J. 63(3), 367-373.

Panfilov, A. V., Pertsov, A. M. (1984) Vortex Ring in Three-Dimensional Active Medium Described by Reaction Diffusion Equation. Dokl. Akad. Nauk. USSR 274(6), 1500-1503 (in Russian).

Panfilov, A. V., Winfree, A. T. (1985) Twisted Scroll Rings in Active Three-Dimensional Media. Physica 17D, 323-330.

Panfilov, A. V., Rudenko, A. N., Pertsov, A. M. (1984) Twisted Scroll Waves in Active Three-Dimensional Media (in Russian). Dokl. Acad. Nauk. USSR 279(4), 1000-1002.

Panfilov, A. V., Rudenko, A. N., Winfree, A. T. (1985) Twisted Vortex Waves in Three-Dimensional Active Media (in Russian). Biofizika 30(3), 464-466.

Paydarfar, D., Eldridge, F. L., Kiley, J. P. (1986) Resetting of Mammalian Respiratory Rhythm: Existence of a Phase Singularity. Am. J. Physiol. 250, R721-R727.

Penrose, R. (1979) The Topology of Ridge Systems. Ann. Hum. Genet. Lond. 42, 435-444.

Perkel, D. H., Schulman, J. H., Bullock, T. H., Moore, G. P., Segundo, J. P. (1964) Pacemaker Neurons: Effect of Regularly Spaced Synaptic Input. Science 145, 61-63.

Pertsov, A. M., Panfilov, A. V., Medvedeva, F. U. (1983) Instabilities of Autowaves in Excitable Media Associated with the Critical Curvature Phenomenon (in Russian). Biofizika 28(1), 100-102.

Pertsov, A. M., Ermakova, E. A., Panfilov, A. V. (1984) Rotating Spiral Waves in a Modified FitzHugh-Nagumo Model. Physica D 14(1), 117-124.

Peterson, E. L. (1980) Phase-Resetting a Mosquito Circadian Oscillator. J. Comp. Physiol. 138, 201-211.

Peterson, E. L. (1981) Dynamic Response of a Circadian Pacemaker. I. Recovery from Extended Light Exposure. II. Recovery from Light-Pulse Perturbations. Biol. Cybern. 40, 171-194.

Peterson, E. L., Calabrese, L. (1982) Dynamic Analysis of a Rhythmic Neural Circuit in the Leech Hirudo Medicinalis. J. Neurophysiol. 47, 256-271.

Poston, T., Winfree, A. T. (1987) Seeds for Organizing Centers.

Randall, W. C., ed. (1977) Neural Regulation of the Heart. New York: Oxford Univ. Press.

Randall, W. C., Thomas, J. X., Euler, D. E., Rozanski, G. J. (1978) Cardiac Dysrhythmias Associated with Autonomic Nervous System Imbalance in the Conscious Dog. In: Neural Mechanisms in Cardiac Arrhythmias, edited by P. J. Schwartz et al. New York: Raven Press.

Rebbi, C. (1979) Solitons. Sci. Am. 235(1), 92-116.

Reiner, V. S., Antzelevitch, C. (1985) Phase-Resetting and Annihilation in a Mathematical Model of the Sinus Node. Am. J. Physiol. 249, H1143-H1153.

Reiner, V. S., Winfree, A. T. (1985) Unpublished computations.

Reshodko, L. B. (1973) Automata Models and Machine Experiments in the Investigation of Biological Systems (e.g. Smooth Muscles). J. Gen. Biol. 1, 80-87.

Reshodko, L. V., Bures, J. (1975) Computer Simulation of Reverberating Spreading Depression in a Network of Cell Automata. Biol. Cybern. 18, 181-189.

Richards, D. A., Blake, G. J., Spear, J. F., Moore, E. N. (1984) Electrophysiologic Substrate for Ventricular Tachycardia: Correlation of Properties in vivo and in vitro. Circ. 69(2), 369-381.

Richter, C. P. (1957) On the Phenomenon of Sudden Death in Animals and Man. Psychosomatic Med. 65, 191-198.

Roberts, D. E., Hersh, L. T., Scher, A. M. (1979) Influence of Cardiac Fiber Orientation on Wavefront Voltage, Conduction Velocity, and Tissue Resistivity in the Dog. Circ. Res. 44, 701-712.

Roelandt, J., Klootwijk, P., Kubsen, J., Janse, M. J. (1984) Sudden Death during Longterm Ambulatory Monitoring. Eur. Heart J. 5, 7-20.

Rolfsen, D. (1976) Knots and Links. Berkeley, Calif.: Publish or Perish Press, pp. 323-416.

Rosenthal, J. E., Ferrier, G. R. (1983) Contribution of Variable Entrance and Exit Block in Protected Foci to Arrhythmogenesis in Isolated Ventricular Tissues. Circ. 67(1), 1-8.

Rossler, O., Kahlert, C. (1979) Winfree Meandering in a 2-Dimensional 2-Variable Excitable Medium. Zeit. Naturforsch. 34, 565-570.

Roy, O. Z., Park, G. C., Scott, J. R. (1977) Intracardiac Catheter Fibrillation Thresholds as a Function of the Duration of 60Hz Current and Electrode Area. IEEE Trans. Biomed. Eng. 24(5), 430-435.

Rozenshtraukh, L., Kholopov, A. V., Yushmanova, A. V. (1970) Vagus Inhibition-Cause of Formation of Closed Pathways of Conduction of Excitation in the Auricles (in Russian). Biofizika 15, 690-700.

Rudenko, A. N., Panfilov, A. V. (1983) Drift and Interaction of Vortices in a Two-Dimensional Active Medium (in Russian). Stud. Biophys. 98, 183-188.

Ruffy, R., Friday, K. J., Southworth, W. F. (1983) Termination of Ventricular Tachycardia by Single Extrastimulation during the Ventricular Effective Refractory Period. Circ. 67(2), 457-459.

Ruoff, P. (1984) Phase Response Relationships of the Closed Bromide-Perturbed Belousov-Zhabotinsky Reaction of the Free Oscillating State. J. Phys. Chem. 88(13), 2851-2857.

Sano, T., Sawanobori, T., Adaniya, H. (1978) Mechanism of Rhythm Determination among Pacemaker Cells of the Mammalian Sinus Node. Am. Phys. Soc. 235, H379-H384.

Sawanobori, T., Hirano, Y., Hirota, A., Fujii, S. (1984) Circus-Movement Tachycardia in Frog Atrium Monitored by Voltage-Sensitive Dyes. Am. J. Physiol. 247, H185-H194.

Schwab, W. E., Josephson, R. K. (1982) Lability of Conduction Velocity during Repetitive Activation of an Excitable Epithelium. J. Exp. Biol. 98, 175-193.

Schwarz, K. W. (1978) Turbulence in Superfluid Helium: Steady Homogeneous Counterflow. Phys. Rev. B. 18, 245-262.

Schwarz, K. W. (1982) Generation of Superfluid Turbulence Deduced from Simple Dynamic Rules. Phys. Rev. Lett. 49, 283-285.

Schwarz, K. W. (1983) Critical Velocity for a Self-Sustaining Vortex Tangle in Superfluid Helium. Phys. Rev. Lett. 50(5), 364-367.

Schwarz, K. W. (1985) Three-dimensional Vortex Dynamics in Superfluid 4He: Line-line and Line-boundary Interactions. Phys. Rev. B 31(9), 5782-5804.

Scott, A. C. (1975) The Electrophysics of a Nerve Fiber. Rev. Mod. Physics 47(2), 487-533.

Scott, S. W. (1979) Stimulation Simulations of Young Yet Cultured Beating Hearts. Ph.D. thesis, State Univ. N.Y. Buffalo.

Selfridge, O. (1948) Studies on Flutter and Fibrillation. V. Some Notes on the Theory of Flutter. Arch. Inst. Cardiol. Mex. 18, 177-187.

Sevcikova, H., Suchanova, D., Marek, M. (1982) The Patterns of Phase Resetting of Oscillation in Zhabotinski Reaction. Scientific Papers Prague Inst. Chem. Tech. (Chemical Engineering) K17, 137-150.

Shcherbunov, A. I., Kukushkin, N. I., Sakson, M. Y. (1973) Reverberator in a System of Interrelated Fibres Described by the Noble Equation (in Russian). Biofizika 18, 519-525.

Shepherd, J. T. (1985) The Heart as a Sensory Organ. J. Am. Coll. Card. 5(6), 83B-87B.

Shibata, N., Chen, P.-S., Summers, E., Wolf, P., Daniely, N. D., Spoon, J. D., Smith, W., Ideker, R. (1985a) Epicardial Activation after Shocks in the Vulnerable Period and in Ventricular Fibrillation. Circ. 72 Suppl. III, 958.

Shibata, N., Chen, P.-S., Worley, S. J., Summers, E., Stilwell, D. J., Ideker, R. E. (1985b) Shock Strength and Ventricular Vulnerability. Circ. 72 Suppl. III, 1528.

Shibata, N., Chen, P.-S., Ideker, R. E. (1986) The Initiation of Ventricular Fibrillation by Stimulation during the Vulnerable Period. Clin. Prog. Electrophysiol. Pacing. In press.

Skinner, J. E. (1985) Regulation of Cardiac Vulnerability by the Cerebral Defense System. J. Am. Coll. Card. 5(6), 88B-94B.

Smith, J. M., Cohen, R. J. (1984) Simple Finite-Element Model Accounts for Wide Range of Cardiac Dysrhythmias. Proc. Natl. Acad. Sci. USA 81, 233-237.

Smith, E. E., Guyton, A. C. (1961) An Iron Heart Model for Study of Cardiac Impulse Transmission. Physiologist 4, 112.

Smith, W. M., Funk, A. L., Ideker, R. E., Bartram, F. R., Talbert, P. V. (1982) A Microcomputer-based Multichannel Data Acquisition System for the Study of Complex Arrhythmias. Computers in Cardiol. 1982, 131-134.

Spach, M. S., Dolber, P. C. (1985) The Relation between Discontinuous Propagation in Anisotropic Cardiac Muscle and the Vulnerable Period of Reentry. In: Cardiac Electrophysiology and Arrhythmias, ed. D. P. Zipes, J. Jalife, pp. 241-252. Orlando, Fla.: Grune & Stratton.

Spach, M. S., Kootsey, J. M. (1983) The Nature of Electrical Propagation in Cardiac Muscle. Am. J. Physiol. 244, H3-H22.

Spach, M. S., Miller, W. T., Geselowitz, D. B., Barr, R. C., Kootsey, J. M., Johnson, E. A. (1981) The Discontinuous Nature of Propagation in Normal Canine Cardiac Muscle. Circ. Res. 48, 39-54.

Spach, M. S., Miller, W. T., Dolber, P. Kootsey, J. M., Sommer, J.,

Mosher, C. (1982) The Functional Role of Structural Complexities in the Propagation of Depolarization in the Atrium of the Dog. Circ. Res. 50, 175-191.

Spielman, S. R., Michelson, E. L., Horowitz, L. N., Spear, J. F., Moore, E. N. (1978) The Limitations of Epicardial Mapping as a Guide to the Surgical Therapy of Ventricular Tachycardia. Circ. 57(4), 666-670.

Stark, L. (1959) Stability, Oscillations, and Noise in the Human Pupil Servomechanism. Proc. IRE, Novem. 1925-1939.

Stephenson, H. (1974) Cardiac Arrest and Resuscitation. St. Louis: Mosby.

Strittmatter, W., Honerkamp, J. (1984) Fibrillation of a Cardiac Region and the Tachycardia Mode of a Two-Oscillator System. J. Math. Biol. 20(2), 171-184.

Strogatz, S. H. (1980) The Mathematics of Supercoiled DNA: An Essay in Geometric Biology. Junior thesis, Princeton University Mathematics Department.

Strogatz, S. H. (1983) The Topology of Zigzag Chromatin. J. Theor. Biol. 103(4), 601-607.

Strogatz, S. H. (1984) Yeast Oscillations, Belousov-Zhabotinsky Waves, and the Non-Retraction Theorem. Math Intelligencer 7(2), 9-17.

Surawicz, B., Zumino, A. P. (1966) The Vulnerable Period of Ventricular Excitation. *In*: Mechanisms and Therapy of Cardiac Arrhythmias, ed. L. S. Dreifus, W. Likoff, J. H. Moyer. New York: Grune & Stratton.

Suzuki, R. (1976) Electrochemical Neuron Model. Adv. Biophys. 9, 115-156.

Suzuki, R., Sato. S., Nagumo, J. (1963) Electrochemical Active Network (in Japanese). Notes of IECE Japan Professional Group on Non-Linear Theory, 26 Feb. 1963.

Swindale, N. V. (1982) A Model for the Formation of Orientation Columns. Proc. Roy. Soc. Lond. B. 215, 211-230.

Swinney, H. L. (1985) Observations of Complex Dynamics and Chaos. *In*: Fundamental Problems in Statistical Mechanics, ed. E.G.D. Cohen. New York: Elsevier Press.

Tabak, V. Y., Chernysh, A. M., Nemirko, A. P., Manilo, L. A. (1980) The Dynamics of Spectral Characteristics of ECG in Cardiac Fibrillation of the Ventricles (in Russian). Anesteziologiia I Reanimatologiia. Moscow 1, 71-79.

Tacker, W. A., Geddes, A. (1980) Fibrillation and Defibrillation. *In*: Electrical Defibrillation, pp. 1-18. Boca Raton, Fla.: CRC Press.

Tavel, M. E., Fisch, C. (1964) Repetitive Ventricular Arrhythmia Resulting from Artificial Internal Pacemaker. Circ. 30, 493-500.

Taylor, W., Krasnow, R., Dunlap, J. C., Broda, H., Hastings, J. W. (1982) Critical Pulses of Anisomycin Drive the Circadian Oscillator in Gonyaulax toward Its Singularity. J. Comp. Physiol. 148, 11-25.

Thom, R. (1972) Structural Stability and Morphogenesis (tr. from French). Reading, Mass.: Benjamin.

Thompson, J.M.T., Hunt, G. W. (1977) The Instability of Evolving Systems. Interdisc. Sci. Rev. 2(3), 240-262.

Thomson, J. (1883) A Treatise on the Motion of Vortex Rings. London: Macmillian.

Thomson, W. I. (1867) On Vortex Atoms. Phil. Mag. 34, 15-24.

Thurston, W. P., Weeks, J. R. (1984) The Mathematics of Three-Dimensional Manifolds. Sci. Am. 251(1), 108-120.

Tomchik, K. J., Devreotes, P. N. (1981) Adenosine 3',5'-Monophosphate Waves in Dictyostelium Discoideum: A Demonstration by Isotope Dilution-Fluorography. Science 212, 443-446.

Troy, W. C. (1978) Mathematical Modelling of Excitable Media in Neurobiology and Chemistry. In: Theoretical Chemistry 4, 133-157. Ed. H. Eyring, D. Henderson. New York: Academic.

Tucker, A. W., Bailey, H. S. (1950) Topology. Sci. Am. 182(1), 18-24.

Turing, A. M. (1952) The Chemical Basis of Morphogenesis. Phil. Trans. Roy. Soc. B237, 37-72.

van Capelle, F. J., Durrer, D. (1980) Computer Simulation of Arrhythmias in a Network of Coupled Excitable Elements. Circ. Res. 47, 454-466.

van Meerwijk, W.P.M, Debruin, G., Ginneken, A. C., van Hartevelt, J., Jongsma, H. J., Scott, S. W. (1984) Phase Resetting Properties of Cardiac Pacemaker Cells. J. Gen. Physiol. 83, 613-629.

Verrier, R. L., Hagestad, E. L. (1985) Role of the Autonomic Nervous System in Sudden Death. In: Sudden Cardiac Death, ed. M. E. Josephson. Philadelphia: F. A. Davis Co.

Vilenkin, A. (1981) Cosmic Strings. Phys. Rev. D 24(8), 2082-2089.

Vilenkin, A. (1984) Formation and Evolution of Cosmic Strings. Phys. Rev. D 30(10), 2036-2045.

Vilenkin, A. (1985) Cosmic Strings and Domain Walls. Phys. Reports 121(5), 263-315.

Voorhees, W. D., Foster, K. S., Geddes, L. A., Babbs, C. F. (1984) Safety Factor for Precordial Pacing: Minimum Current Thresholds for Pacing and for Ventricular Fibrillation by Vulnerable-Period Stimulation. Pace 7, 356-360.

Wang, J. C. (1982) DNA Topoisomerases. Sci. Am. 247(1), 94-109.

Wasserman, S. A., Dungan, J. M., Cozzarelli, N. R. (1985) Discovery of a Predicted DNA Knot Substantiates a Model for Site-specific Recombination. Science 229, 171-174.

Wellens, H.J.J., Vermeulen, A., Durrer, D. (1972) Ventricular Fibrillation Occurring on Arousal from Sleep by Auditory Stimuli. Circ. 46, 661-664.

Welsh, B. J. (1984) Pattern Formation in the Belousov-Zhabotinsky Reaction. Ph.D. thesis, Glasgow College of Technology.

Welsh, B., Gomatam, J. (1984) A Trial Phase Function Method for a Class of Lambda-Omega Reaction-Diffusion Systems: Reduction to a Schrodinger Type Problem in an Eigen Sub-Domain. Adv. Appl. Ma. 5(3), 333-355.

Welsh, B., Gomatam, J., Burgess, A. (1983) Three-Dimensional Chemical Waves in the Belousov-Zhabotinsky Reaction. Nature 304, 611-614.

West, T. C., Landa, J. F. (1962) Minimal Mass Required for Induction of a Sustained Arrhythmia in Isolated Atrial Segments. Am. J. Physiol. 202, 232-236.

Wever, R. (1962) Zum Mechanismus der Biologischen 24-Stunden-Periodik I. Kybernetik 1(1), 139-154.

Wever, R. (1963) Zum Mechanismus der Biologischen 24-Stunden-Periodik II. Kybernetik 1(6), 213-231.

Wever, R. (1964) Zum Mechanismus der Biologischen 24-Stunden-Periodik III. Kybernetik 2(3), 127-144.

Wheeler, J. (1964) Geometrodynamics and the Issue of the Final State. In: Relativity, Groups, and Topology, ed. C. DeWitt, B. S. DeWitt. New York: Gordon and Breach.

Wiedenmann, G. (1977) Weak and Strong Phase Shifting in the Activity Rhythm of Leuophaea Maderae (Blaberidae) after Light Pulses of High Intensity. Zeit. Naturforsch. 32c, 464-465.

Wiener, N., Rosenblueth, A. (1946) The Mathematical Formulation of the Problem of Conduction of Impulses in a Network of Connected Excitable Elements, Specifically in Cardiac Muscle. Arch. Inst. Cardiol. Mex. 16, 205-265.

Wiggers, C. J., Wegria, R. (1939) Ventricular Fibrillation due to Single Localized Induction and Condenser Shocks Applied during the Vulnerable Phase of Ventricular Systole. Am. J. Physiol. 128, 500-505.

Wilson, H. R., Cowan, J. D. (1972) Excitatory and Inhibitory Interactions in Localized Populations of Model Neurons. Biophysical J. 12, 1-24.

Winfree, A. T. (1967) Biological Rhythms and the Behavior of Populations of Coupled Oscillators. J. Theor. Biol. 16, 15-42.

Winfree, A. T. (1968) The Investigation of Oscillatory Processes by Perturbation Experiments. in: Biol. Biochem. Oscillators, ed. B. Chance, E. K. Pye, B. Hess, pp. 461-501. London: Academic Press.

Winfree, A. T. (1970a) An Integrated View of the Resetting of a Circadian Clock. J. Theor. Biol. 28, 327-374.

Winfree, A. T. (1970b) Oscillatory Control of Differentiation in Nectria. Proc. of IEEE Adaptive Systems Symposium, Dallas, Tex., 23, 4.1-4.7.

Winfree, A. T. (1970c) The Temporal Morphology of a Biological Clock. Lectures on Mathematics in the Life Sciences 2, 109-150, ed. M. Gerstenhaber. Providence, R.I.: A.M.S.

Winfree, A. T. (1970d) 16mm movie for diverse symposia.

Winfree, A. T. (1971) Spiral Chemistry. Forskning och Framsteg 6, 9-10 (in Swedish).

Winfree, A. T. (1972a) Spiral Waves of Chemical Activity. Science 175, 634-636.

Winfree, A. T. (1972b) On the Photosensitivity of the Circadian Time-Sense in Drosophila. J. Theor. Biol. 35, 159-189.

Winfree, A. T. (1972c) Oscillatory Glycolysis in Yeast: The Pattern of Phase Resetting by Oxygen. Archives of Biochem. and Biophys. 149, 338-401.

Winfree, A. T. (1973a) Polymorphic Pattern Formation in the Fungus Nectria. J. Theor. Biol. 38, 362-382.

Winfree, A. T. (1973b) Scroll-Shaped Waves of Chemical Activity in Three Dimensions. Science 181, 937-939.

Winfree, A. T. (1973c) Spatial and Temporal Organization in the Zhabotinsky Reaction. Adv. Biol. Med. Phys. 16, 115-136.

Winfree, A. T. (1973d) Resetting the Amplitude of Drosophila's Circadian Chronometer. J. Comp. Physiol. 85, 105-160.

Winfree, A. T. (1974a) Rotating Chemical Reactions. Sci. Am. 230(6), 82-95.

Winfree, A. T. (1974b) Rotating Solutions to Reaction/Diffusion Equations. S.I.A.M./A.M.S. Proc. 8, 13-31, ed. D. Cohen.

Winfree, A. T. (1974c) Wavelike Activity in Biological and Chemical Media. Lecture Notes on Biomathematics 2, 243-260, ed. P. van den Driessche.

Winfree, A. T. (1974d) Two Kinds of Wave in an Oscillating Chemical Solution. Faraday Symp. Chem. Soc. 9, 38-46.

Winfree, A. T. (1974e) Patterns of Phase Compromise in Biological Cycles. J. Math. Biol. 1, 73-95.

Winfree, A. T. (1975a) Unclocklike Behavior of a Biological Clock. Nature 253, 315-319.

Winfree, A. T. (1975b) Resetting Biological Clocks. Physics Today 28, 34-39.

Winfree, A. T. (1976) On Phase-Resetting in Multicellular Clockshops. *In*: The Molecular Basis of Circadian Rhythms, ed. W. Hastings, H. G. Schweiger, pp. 109-129. Berlin: Abakon.

Winfree, A. T. (1977) The Phase Control of Neural Pacemakers. Science 197, 761-762.

Winfree, A. T. (1978) Stably Rotating Patterns of Reaction and Diffusion. Prog. Theor. Chem. 4, 1-51, ed. H. Eyring, D. Henderson.

Winfree, A. T. (1979) 24 Hard Problems about 24-Hour Rhythms. *In*: Non-Linear Oscillations in Biology, Lectures in Applied Math. 17, ed. F. Hoppensteadt. Providence, R.I.: A.M.S.

Winfree, A. T. (1980) The Geometry of Biological Time. New York: Springer-Verlag.

Winfree, A. T. (1981) Peculiarities in the Impulse Response of Pacemaker Neurons. *In*: Math. Aspects of Physiol., Lectures in Applied Math. 19, 265-279, ed. F. Hoppensteadt. Providence, R.I.: A.M.S.

Winfree, A. T. (1982a) Fibrillation as a Consequence of Pacemaker Phase-Resetting. *In*: Cardiac Rate and Rhythm, ed. L. N. Bouman, H. J. Jongsma, pp. 447-472. The Hague: M. Nijhoff.

Winfree, A. T. (1982b) The Rotor as a Phase Singularity of Reaction-Diffusion Problems and Its Possible Role in Sudden Cardiac Death. *In*: Non-Linear Phenomena in Chemical Dynamics, ed. C. Vidal, A. Pacault, pp. 156-159. Berlin: Springer-Verlag.

Winfree, A. T. (1983) Sudden Cardiac Death—A Problem in Topology. Sci. Am. 248(5), 144-161.

Winfree, A. T. (1984a) The Prehistory of the Belousov-Zhabotinsky Reaction. J. Chem. Educ. 61, 661-663.

Winfree, A. T. (1984b) Wavefront Geometry in Excitable Media: Organizing Centers. Physica 12D, 321-332.

Winfree, A. T. (1984c) A Continuity Principle for Regeneration. *In*: Pattern Formation, ed. G. Malacinski. New York: Macmillan.

Winfree, A. T. (1985) Organizing Centers for Chemical Waves in Two and Three Dimensions. *In*: Oscillations and Traveling Waves in Chemical Systems, ed. R. Field, M. Burger, pp. 441-472. New York: Wiley.

Winfree, A. T. (1986a) The Timing of Biological Clocks. Scientific American Library.

Winfree, A. T. (1986b) Filaments of Nothingness. The Sciences, Mar./Apr. 20-27.

Winfree, A. T., Strogatz, S. H. (1983a) Singular Filaments Organize Chemical Waves in Three Dimensions: 1. Geometrically Simple Waves. Physica 8D, 35-49.

Winfree, A. T., Strogatz, S. H. (1983b) Singular Filaments Organize Chemical Waves in Three Dimensions: 2. Twisted Waves. Physica 9D, 65-80.

Winfree, A. T., Strogatz, S. H. (1983c) Singular Filaments Organize

Chemical Waves in Three Dimensions: 3. Knotted Waves. Physica 9D, 333-345.

Winfree, A. T., Strogatz, S. H. (1984a) Singular Filaments Organize Chemical Waves in Three Dimensions: 4. Wave Taxonomy. Physica 13D, 221-233.

Winfree, A. T., Strogatz, S. H. (1984b) Organizing Centers for Three-Dimensional Chemical Waves. Nature 311(5987), 611-615.

Winfree, A. T., Twaddle, G. (1981) The Neurospora Mycelium as a 2-Dimensional Sheet of Coupled Circadian Clocks. *In*: Mathematical Biology, ed. T. A. Burton. New York: Pergamon Press.

Winfree, A. T., Winfree, E. M., Seifert, H. (1985) Organizing Centers in a Cellular Excitable Medium. Physica 17D, 109-115.

Wit, A. L., Cranefield, P. F. (1976) Triggered and Automatic Activity in the Canine Coronary Sinus. Circ. Res. 41(4), 435-445.

Wit, A. L., Cranefield, P. F. (1978) Reentrant Excitation as a Cause of Cardiac Arrhythmias. Am. J. Physiol. 235(1), H1-H17.

Wit, A. L., Rosen, M. R. (1981) Cellular Electrophysiology of Cardiac Arrhythmias, Part II: Arrhythmias Caused by Abnormal Impulse Conduction. Modern Concepts of Cardiovascular Disease 50, 7-12.

Wit, A. L., Hoffman, B. F., Rosen, M. R. (1975) Electrophysiology and Pharmacology of Cardiac Arrhythmias. I_x Effects of Beta Adrenergic Receptor Stimulation and Blockade. Am. Heart J. 90, 521-533.

Wit, A. L., Allessie, M. A., Bonke, F.I.M., Lammers, W., Smeets, J., Febifkui, J. J. (1982) Electrophysiologic Mapping to Determine the Mechanism of Experimental Ventricular Tachycardia Initiated by Premature Impulses. Am. J. Cardiol. 49, 166-185.

Witkowski, F. X., Corr, P. B. (1984) An Automated Simultaneous Transmural Cardiac Mapping System. Am. J. Physiol. 247, H661-H668.

Wright, F. J., Nye, J. F. (1982) Dislocations in Diffraction Patterns: Continuous Waves and Pulses. Phil. Trans. Roy. Soc. Lond. A 305, 339-382.

Yakushevich, L. V. (1984) Vortex Filament Elasticity in Active Medium. Studia Biophysica 100(3), 195-200.

Yanagihara, K., Noma, A., Irisawa, H. (1980) Reconstruction of Sino-Atrial E Pacemaker Potential Based on the Voltage Clamp Experiments. Jpn. J. Physiol. 30(6), 841-857.

Zaikin, A. N., Zhabotinsky, A. M. (1970) Concentration Wave Propagation in Two-Dimensional Liquid-Phase Self-Oscillating Systems. Nature 225, 535-537.

Zeeman, E. C. (1977) Catastrophe Theory: Selected Papers. Reading, Mass.: Addison-Wesley.

Zel'dovich, Y. B., Sokolov, D. D. (1984) Zeros of Eigenfunctions of Free and Forced Oscillations of Continuous Systems with Dissipation. Dokl. Akad. Nauk. USSR 275(2), 358-361.

Zhabotinsky, A. M. (1964) Periodic Movement in the Oxidation of Malonic Acid in Solution (Investigation of the Kinetics of Belousov's Reaction) (in Russian). Biofizika 9(3), 306-311.

Zhabotinsky, A. M. (1970) Investigations of Homogeneous Chemical Auto-Oscillating Systems (in Russian). Thesis, Puschino USSR.

Zhabotinsky, A. M. (1985) The First Period 1961-1969 of Systematic Studies of Oscillations and Waves in Belousov Chemical Systems. *In*: Oscillations and Traveling Waves in Chemical Systems, ed. R. Field, M. Burger. New York: Wiley.

Zipes, D. P. (1975) Electrophysiological Mechanisms Involved in Ventricular Fibrillation. Circ. Supp. III to vols. 51 & 52, III-120 to III-130.

Zipes, D. P., Heger, J. J., Miles, W. M., Mahomed, Y., Brown, J. W., Spielman, S. R., Prystowsky, E. N. (1984) Early Experience with an Implantable Cardioverter. N. Engl. J. Med. 311(8), 485-490.

Zykov, V. S. (1980) Analytical Estimation of Dependence of Excitation Wave Velocity in Two-Dimensional Excitable Medium on Curvature of Its Front (in Russian). Biofizika 25(5), 888-892.

Zykov, V. S. (1984) Modelling of Wave Processes in Excitable Media (in Russian). Moscow: Nauka. (Eng. trans. in prep., ed. A. T. Winfree, forthcoming Manchester Univ. Press.)

Zykov, V. S., Morozova, O. L. (1979) Velocity of the Excitation Wave Propagation in a Two-Dimensional Excitable Medium (in Russian). Biofizika 23, 717-722.

Abakumov et al. (1970), 109
Abildgaard (1775), 138
Adgey et al. (1982), 47
Agladze et al. (1983), 119, 152, 180, 181, 182, 184
Alessi et al. (1958), 111
Allessie and Bonke (1978), 134
Allessie et al. (1973), 113, 134
Allessie et al. (1976), 101, 113, 114, 115, 134, 272
Allessie et al. (1977), 114, 119, 134
Allessie et al. (1982), 117, 118, 122
Allessie et al. (1984), 115, 117, 120, 134
Allessie et al. (1985), 105, 112, 117, 120, 132, 134, 156, 157
Anderson (1980), 120
Antzelevitch and Moe (1983), 150
Armour and Sinha (1975), 111
Armour et al. (1972), 46, 111

Balakhovsky (1965), 108, 133
Baranova et al. (1983), 200
Bardy et al. (1983), 113, 127
Battersby (1965), 112
Bauer et al. (1979), 225
Beeler and Reuter (1977), 94, 236
Bellet (1971), 39, 40, 101, 140
Belousov (1959), 161
Belousov (1985), 13
Bernal (1965), 154
Bernstein and Phillips (1981), 243, 260
Berry (1980), 200
Berry (1981), 200
Best (1976), 68, 74, 75, 76, 77, 79, 81
Best (1979), 74, 75, 76, 77, 79, 132, 268
Bleeker (1982), 127
Boineau et al. (1978), 127
Boineau et al. (1980a), 101, 113, 120
Boineau et al. (1980b), 127
Bolker (1973), 260
Bonner (1967), 173
Botre et al. (1983), 185
Bouligand (1972), 258
Bray (1921), 154
Briller (1966), 140

Brooks et al. (1955), 140, 142
Brown and Eccles (1934a), 53, 88
Brown and Eccles (1934b), 53, 88
Brown et al. (1984a), 88
Brown et al. (1984b), 88
Brusca and Rosettani (1973), 103, 120, 211
Bryant et al. (1977), 257
Bryant et al. (1981), 257
Buck (1973), 120
Burch and Depasquale (1965), 46
Busse (1969), 166

Cardinal et al. (1984), 120
Castellanos et al. (1984), 93
Chay and Lee (1984), 88, 94, 100, 132, 150, 275
Chay and Lee (1985), 94, 100, 132, 150, 275
Chen et al. (1986a), 135, 139
Chen et al. (1986b), 139
Clay et al. (1984), 58, 275
Cohen et al. (1978), 279
Cohn et al. (1982), 213
Cole (1968), 155
Colizza et al. (1983), 90
Corr and Gillis (1978), 46
Cranefield et al. (1957), 142
Crick (1976), 225

Daggett and Wallace (1966), 144
de Bakker et al. (1979), 134, 211
de Bakker et al. (1983), 120
de Bruin et al. (1984), 113
de Vega (1978), 243
Delbruck (1962), 225
Denes et al. (1981), 100
DeSilva (1982), 46
DeSimone et al. (1973), 192
Dharmananda (1980), 27
Dharmananda and Feldman (1979), 27
Diaz et al. (1984), 88
Dillon and Morad (1981), 121
Dillon et al. (1985), 135
Dolnik et al. (1984), 164

LIBRARY OF CONGRESS CATALOGING-IN-PUBLICATION DATA

Winfree, Arthur T.
When time breaks down.

Bibliography: p.   Includes indexes.
1. Biological rhythms.   2. Arrhythmia.   3. Space and time.   I. Title.
QP84.6.W56 1986   612'.022   86-12363
ISBN 0-691-08443-2 (alk. paper)
ISBN 0-691-02402-2 (pbk.)

Arthur T. Winfree is Professor in the Department of Ecology and Evolutionary Biology and the
Program in Applied Mathematics at the University of Arizona (Tucson, Arizona 85721).